THE STONES OF ATLANTIS

NEW AND REVISED

Dr. David Zink

New York London Toronto Sydney Tokyo Singapore

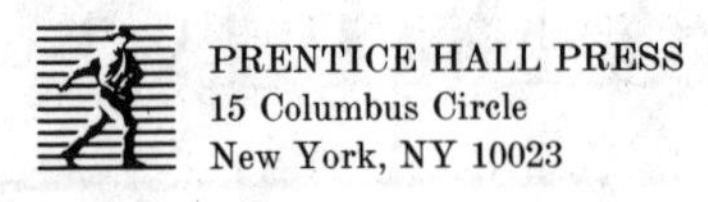

PRENTICE HALL PRESS
15 Columbus Circle
New York, NY 10023

Library of Congress Cataloging-in-Publication Data
Zink, David, 1927–
The stones of Atlantis / David D. Zink.—New and rev.
p. cm.
Includes bibliographical references.
ISBN 0-13-847096-0
1. Atlantis. 2. Interplanetary voyages. 3. Bimini Islands
(Bahamas)—Antiquities. 4. Bahamas—Antiquities. I. Title.
GN751.Z57 1990
917.296—dc20 90-33243
CIP

Designed by Richard Oriolo

Manufactured in the United States of America

10 9 8 7 6 5 4 3 2 1

Revised Edition

ACKNOWLEDGMENTS

The field exploration at Bimini on which the present work is based would have been impossible without the generous participation of many people. Inevitably, not all can be acknowledged by name. I wish to offer my warm personal thanks to those not here mentioned, especially expedition members. I am particularly indebted to the following:

Larry E. Arnold, Frank Auman, William Beidler, Ph.D.; Dr. William Bell, Neville, Ossie, and Harcourt Brown, Edgar Evans Cayce, Gail Cayce, Hugh Lynn Cayce, C. W. Conn, Jr.; John DeGarmo, Rick Frehsee, Karen Getsla, Dick and Buffy Hart, Carol Huffstickler, Linda E. Larson, Roy Lemon, Ph.D.; Joseph Libbey, Jerry Long, Raymond McAllister, Ph.D.; John Parks, Douglas G. Richards, John O. Sherman, Jr., Capt. USN (Ret.); Jon Douglas Singer, Frank O. Spampinato, John Jacob Steele, Kiiri Tamm, Peter Tompkins, William Trigg, J. Manson Valentine, Ph.D.; Debbie and Ben Van Der Swan, Marcel Vogel, Ph.D.; John Whitford, Richard Wingate, David Paul Zink, and Laurie Wilson Zink.

My thanks also to Senator Doris L. Johnson, Ed.D., president of the Bahamian Senate and director of the Bahamas Antiquities Institute; the Edgar Cayce Foundation, as well as Dale and Barbara Boden, whose hospitality furthered the writing, and to Ray Jones, Aqua Lung Diving Center, and Gamma Industries. To Claire Gerus, my editor, without whose patience, insight, and long hours of creative editing this book would not have achieved its present form, and to Allan Stormont, my publisher, whose vision brought the manuscript to fruition. My deep appreciation to my long time friend and adviser, Joan.

Notwithstanding the assistance I have been given in various scientific areas, I assume full responsibility for any statement or interpretation that further investigation may render invalid.

ACKNOWLEDGMENTS FOR THE REVISED EDITION

I wish to express my appreciation to the following individuals for their assistance in the pursuit of the Poseidia project from 1977 on:

Alan P. Curtner for nuclear activation analysis of core samples at Virginia Polytechnic Institute and State University; participants in Poseidia '78 and Poseidia '79 who included the late Dr. Harold Edgerton, George Doyle, George W. Hersh, Larry Haley, Paul Evancoe, Terry Mahlman, Janet MacArthur, Charles E. Donnally, Rob Peterman, Stan Janecka, and Roger Crossland; Mark Hosey, who, early in the project, began to contribute to the logistics in many practical ways; Dr. C. Lavett Smith, who provided important insights into the findings of Poseidia '79; Ken Johnson, for his analysis of a possible connection between Balkan prehistory and the marble head found at Bimini Island; Raymond E. Leigh, Jr. for his persistence in identifying the various mounds on East Bimini and for sharing his analysis of the geometric relationships in appendix E; William Donato for his analysis of North American mound sites and ancient measurements worldwide; and Doug Schulkind, my editor, for his cooperative spirit and support in updating this edition.

Finally, I want to thank the following members of the Institute for Planetary Studies for their supportive research and editorial suggestions: Kathleen Law, Antonio Molina, Vanda Osmon, John Dan Reib, and Joan Wilson Zink. Special thanks are due to John Reib for entering the text of the revised edition into a word processor and to Joan Zink for her agreement to go public with the psychic readings she did as "Anne."

CONTENTS

PART III
THE PLEIADES THEORY

PART IV
ON BOARD AGAIN: POSEIDIA '76–'79

FOREWORD TO THE REVISED EDITION

When, over thirty years ago, I began the quest narrated in the present work, it involved for the most part research in literary sources.

At that time, understanding of the physical processes of the earth's past made Plato's story of Atlantis quite improbable. In geology, for example, it is really only in the last decade that Sir Charles Lyell's nineteenth-century uniformitarianism has been seriously challenged by an emerging neo-catastrophism. The result of this evolution is a more realistic awareness of the hazards facing the human race, hazards posed by the ongoing physical evolution of planet Earth.

The scenarios for the destruction of Atlantis offered by Plato and, later, Edgar Cayce, were ignored by those earth scientists committed to some version of Lyell's gradualism. Now it is a different story. My last major expedition to the Bahamas took place a couple of years before this conceptual evolution took place.

The challenge that emerged in organizing the Poseidia expeditions was enlisting the help of scientists whose expertise was directly applicable to expedition objectives —on a controversial site. When Poseidia expeditions produced archaeological anomalies, trained archaeologists were on the staff, one of them a student of George Bass, who (in this country) invented underwater archaeology. The critical lab work was done at nationally known institutions.

Over the years that I have pursued this subject, the parable offered by the story of Atlantis has been taken up independently by those who see the very real possibility that the human race may well destroy the planet that nurtures us. In 1976, when Phillipe Cousteau interviewed me in the Bahamas, he asked, "If the existence of Atlantis could be proven, what would be the meaning for humanity?" My response was that we were presently replaying the script; the satisfaction of our selfish personal needs has taken priority over living in harmony with an ultimately fragile planet.

Fortunately, in the last decade, this concern has begun to be addressed globally. For example, the National Academy of Sciences in this country began last year to meet with Soviet colleagues to consider global atmospheric pollution and its consequences.

Clearly, the events of the past thirty years have worked toward a validation of the Gaia hypothesis advanced by British biologist James Lovelock—essentially the idea that the design of the earth's geosphere and biosphere has yet to be clearly understood, particularly the elements of this design developed before humanity's arrival. In ignorance of this system, humanity has become increasingly destructive of Mother Earth. Somehow ancient peoples understood this better than we presently do. They understood the need to function in harmony with nature. The destruction of Atlantis was one of those events in which humanity's malignant narcisism led to the destruction of all that we value.

All in all, my experience with the research described in the present work has indicated that, from the point of view of science, an objective systems analysis is the only way to get to the truth. This objectivity requires individual scientists to assess their cognitive filters *and* their affective filters honestly in order to be truly objective. For humanity at large, let us hope that everyone, especially those in decision-making positions, understands that continuation of our present lifestyles will destroy the planet.

Finally, it should be clear in the present work that the topic has vital relevance for all humanity and that it cannot be understood clearly through individual projects. Interdisciplinary cooperation is vital to understanding this ancient site.

As someone once said, the scholar's task is "to find the truth no matter how obscure; to recognize it no matter in what strange form it may present itself; to formulate it honestly; to state it unmistakably; and to reason from it remorselessly and without regard to prejudice."

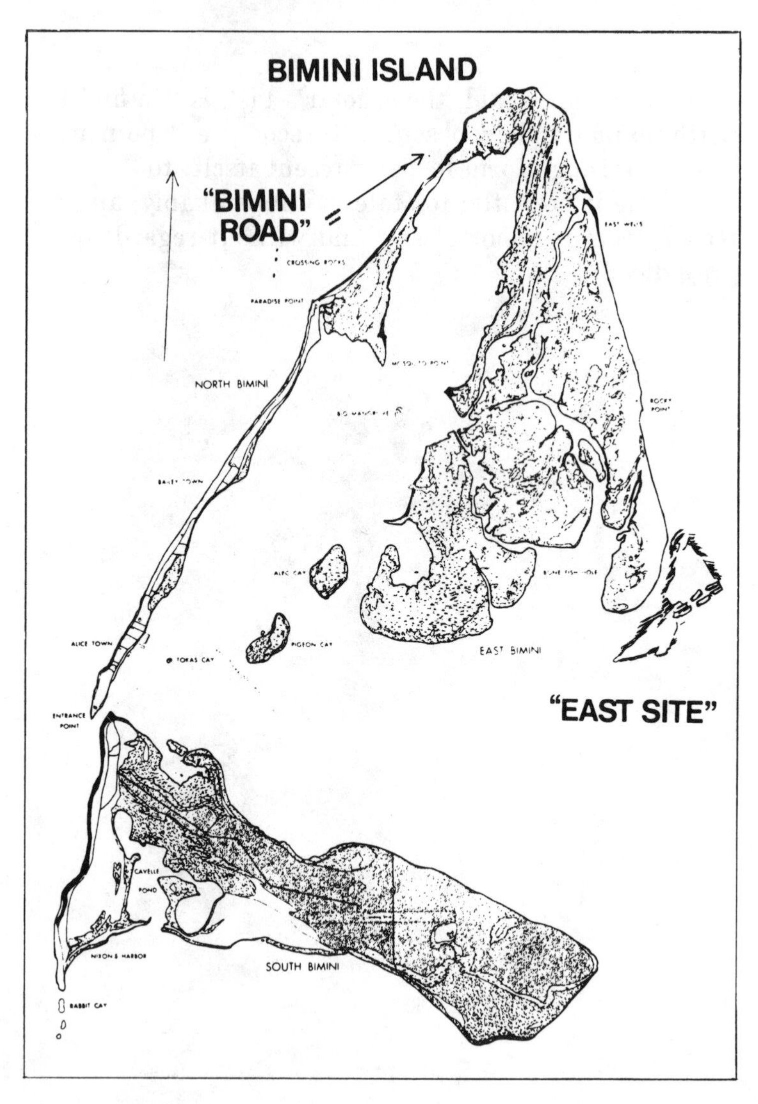

Bimini Island.

INTRODUCTION: THE QUEST BEGINS

Late one evening in 1975 I sat by myself, lost in thought in Brown's Bar on North Bimini Island. It was a sultry June night; the prevailing wind had died. Inside, under the air-conditioning, Neville Brown tidied up the bar while watching television. The bar was quiet because most of the island's visitors and the members of my expedition were elsewhere on the island dancing to live native music. How, I wondered, could an English professor like myself have gotten so far afield from Victorian literature as to pursue Atlantis as a serious archaeological possibility? It was one thing to teach Plato's myths to students of literature and quite another to contend seriously that these tales may have contained some truths, that there may really *have* been an Atlantis. Obviously I believed the latter or I would not have embarked upon this expedition. A decade ago, few people—myself least of all—would have suggested that I would become involved in such a quest. To have anticipated several actual expeditions within the Bermuda Triangle would have been even more farfetched.

Under 17 feet of water, just 3 miles from where I sat in Brown's Bar, rested the mysterious stones that had once again brought me back to the Bahamas. I had returned for the second time more convinced than ever that these huge encrusted limestone blocks, now covered by the sea for at least 6,000 years, might possibly hold the key to an important chapter in prehistory. Were they, as some people insisted, merely natural geological formations? Or could they be something altogether stranger and more significant—perhaps even the remains of Atlantis or one of her colonies? I fervently hoped that the present expedition would penetrate the secrets of these cyclopean stones off Paradise Point. By now, whichever way it went, I wanted some answers.

My background as an Air Force communications and electronics officer was hardly the expected beginning for a mission of this kind. Yet it was in the military service that I learned many of the skills and disciplines that later proved so valuable to my quest. As a commander of communications posts in Korea, I juggled responsibilities such as maintaining generator plants, road building, obtaining (sometimes by serendipity) necessary supplies, and supervising truck operations—all under extremely primitive conditions.

After four years of this and other field assignments, my interest shifted from electronics to teaching, and I won a post at the new Air Force Academy in Denver, Colorado. Sent to graduate school to prepare for the assignment, I had the good luck to take a course under Professor Joseph Cohen, who founded the university honors program in the United States. His seminar on philosophy and literature was a freewheeling "head trip" through the great themes of western culture, drawing upon the sciences as well as the arts to trace their origins.

Once at the academy, I was delighted to find an atmosphere of flexibility and innovation. The stereotypes about "the military mind" certainly did not apply to my fellow officers and faculty members—a lively, enthusiastic group. I was asked to teach a two-semester humanities course structured around classics of Western literature. This course was designed to lure the cadets' attention from the more glamorous space-age courses, so we in the English department worked hard to find the best titles and the most effective presentations. In our study of the Greek tragedies, we did everything possible to give students the feeling of the period. For example, our classes were assigned Robert Graves' *The Greek Myths* to provide them with the biological, anthropological, and political dimensions of the various myths on which the dramas were based. This approach paid off as the cadets became intrigued by the classics, and the course began to be more exciting.

My own students ranged afar in their reading, from Peter Tompkins' *The Virgin and the Eunuch* to Velikovsky's *Worlds in Collision,* and in self-defense I tried to keep up with them. For both faculty and students, this cross-fertilization process greatly enlarged our intellectual horizons and introduced us to some of the surprising new theories about the prehistory of our planet. In the light of my rather conventional graduate education, I was at first inclined to dismiss these ideas as pretty farfetched, but I eventually became fascinated enough with these theories to wonder just how far scholarly and scientific evidence might go toward supporting them. One of the most colorful of the new theorists was Immanuel Velikovsky, who observed that legends and myths around the world involved seven destructions by fire and water. He proposed that these tales represented the residue of terrible natural violence actually endured by prehistoric peoples. He looked for—and appeared to find—scientific evidence of such catastrophes. As triggers to such violence, he particularly emphasized the close passages of Mars and Venus to Earth.

Although Velikovsky's theories were the target of much abuse for many years, a change in scientific attitude occurred in 1973, when he was invited to speak at Harvard. The following year, the highly prestigious American Association for the Advancement of Science held a meeting to consider his theories. A number of discoveries by the United States space program have since borne out several predictions he deduced from his theory. For example, Venus was discovered to be about 800 degrees Fahrenheit—or incandescent, as Velikovsky had predicted. (Einstein, incidentally, had erroneously suggested the planet to be a frigid −13 degrees Fahrenheit.) Jupiter, predicted by Velikovsky to be a source of radio energy, was indeed discovered to be radiating in the 80-megahertz range.

After considering Velikovsky's writings and witnessing his vindication by scientific fact, it became increasingly clear to me that prehistory might well contain a series of mind-shattering earth changes for early humans. These experiences could, as Velikovsky claimed, have been repressed into what Carl Jung would call the collective unconscious or the racial memory. Jung perceived the human unconscious as a layered mechanism—there was a personal unconscious and a deeper layer called the collective unconscious. (Recent psychological research has lent some support to this hypothesis.) The collective unconscious contains universal ideas common to all people, or "archetypes." For example, the recurrent interest in Atlantis could be a result of its existence as an archetypal memory in our unconscious. On the other hand, my respect for the rules of evidence and scientific methods made me skeptical and hungry for actual proof.

As I continued my researches, I became aware that Atlantis was probably only one example of an extinct civilization. Actual human events are often lost to history in three ways: (1) the record is erased by natural destruction of a city or a people; (2) fallible human memory plays a part; (3) evidence is wiped out through the egotistical desire of conquerors to efface the work of all who have gone before. This may also be intensified by religious or political motivations. When archaeological discoveries confirm the existence of these mythical civilizations, the tales themselves take on a new meaning. In the first case, Troy was a fabled city described by Homer until Heinrich Schliemann, a romantic German millionaire, found the site at Hissarlik in Turkey in 1871, decisively removing Troy from literary fantasy. A terrible eruption of Vesuvius in A.D. 79 buried the towns of Pompeii and Herculaneum deep under volcanic ash for nearly 1,700 years. During these many centuries until their excavation, their existence almost vanished from memory. In 1738, excavations revealed the full horror of this natural disaster and brought the event back to life in historical consciousness.

When I was a graduate student, King Arthur and his fabled Knights of the Round Table were thought to be purely literary. Now recent archaeological work in Wales has brought even Arthur into the realm of historical possibility. During my academic career, I watched this process of myth being transformed into history with increasing frequency.

Even the relatively recent past has often escaped us. Not until 1969 did we become aware of an accomplishment of Tudor metallurgy. In 1545, Henry VIII's technicians knew how to make wrought-iron cannons from sheets of metal, including gas-tight welds the length of the barrel. This evidence appeared during a search of the wreckage of the *Mary Rose,* the pride of Henry's fleet, which had rested under the silt of Portsmouth harbor for 400 years. In view of our ignorance of the details of even the last few centuries, eclipses of the facts of history should not be surprising.

In addition to natural catastrophes there is the conqueror's tendency to downgrade the past, evident throughout history. As the paleontologist Loren Eiseley has put it: "Do you know that history is full of evidence of hatred for the past, of a desire on the part of some men to destroy even the memory of their predecessors? Public monuments are effaced, names destroyed, histories rewritten. Sometimes to achieve these ends a whole intellectual elite may be slaughtered in order that the peasantry can be deliberately caused to forget its past." I personally feel that two events in particular have done nearly irreparable damage to the history of the ancient world. One was the destruction of the Alexandrian library: in A.D. 642 the troops of Caliph Omar burned the library's contents to heat the baths of Alexandria for several months. The Caliph is reported to have said, "If what they say is in the Koran, they are useless and you may burn them. If what they say is not in the Koran, they are pernicious and must be destroyed." In view of the incredible sophistication of Egyptian learning (for example, surveyors in the third millenium B.C. could establish longitude and latitude with greater accuracy than would be again known until the eighteenth century A.D.), this book-burning was disastrous for our knowledge of ancient history.

During the Spanish invaders' general destruction of the Mayan and Aztec cultures of Mexico in the sixteenth century, Diego de Landa, later Bishop of Yucatan, burned all but three Mayan writings. These three surviving texts, now located in Paris, Dresden, and Madrid, have so far proved an inadequate base for a full reconstruction of the Mayan language. Thus two acts of institutional religious fanaticism have seriously hampered the search for knowledge of the two cultures, erasing certain vital facts from recorded history.

As a result, I found myself moving away from traditional concepts of myths as primitive, unscientific, inadequate interpretations of natural phenomena. More valid for me was MIT Professor Giorgio de Santillana's understanding of myth as encoded messages containing astronomical and geophysical information. Reading his *Hamlet's Mill,* I was fascinated to discover that one of the most respected historians of science had come to such a sophisticated view. These relics of ancient consciousness, Santillana affirms, contain "the remnants of a preliterate 'code language' of unmistakable coherence . . . a common language which ignored local beliefs and cults . . . concentrated on numbers, motions, measures, overall frames . . . on the structure of numbers, on geometry." He describes Egyptian and Mesopotamian "ritual documents" as "the last form of international initiatic (or esoteric) language, intended to be misunderstood alike by suspicious authorities and the ignorant crowd." Combining Santillana's views with Carl Jung's concept of the unconscious and racial memory, one might argue that myths represent many realities of the past, as Velikovsky and others had claimed.

My newfound interest in myth and prehistory was, increasingly, beginning to focus on the legends of the lost civilization of Atlantis. Like the Troy legend, the Atlantean myth seemed to offer the tantalizing possibility of some sort of basis in fact. Thus, I felt that the discovery of reliable evidence for the existence of Atlantis would dramatize not only the deficiencies in our historical knowledge but also the value of myth and psychology as tools of historical research.

But I was also aware of the intellectual dangers involved in permitting myself to become too much preoccupied with Atlantis. Most academics tend to be wary of the subject and to look with a certain amount of suspicion on those of their colleagues who seem too credulous. Understandably. Across the years, the pursuit of the historical Atlantis has attracted such an extraordinary number of charlatans and cranks that almost any new "evidence" of its existence has to be treated with the greatest caution. I promised myself to do everything in my power to avoid falling victim to the "Atlantis obsession." While I would be prepared to investigate any new evidence and entertain any new hypothesis, no matter how fragmentary, I would always insist on having concrete proof before *accepting* them.

By the same token, I saw no reason to avoid unconventional methods in pursuing the investigation. It made no difference what I might think about the reliability of these methods beforehand; the only worthwhile test was whether or not they eventually produced any verifiable results. Thus, I was perfectly willing at least to consider what paranormal sources had to say on the subject of Atlantis.

During a visit with friends in Boulder, Colorado, at the end of the sixties, I had been introduced to the trance-state material amassed in readings by the famous American clairvoyant Edgar Cayce. My friend Frank Richter, a levelheaded engineer at Dow Chemical, told me of his excitement over the Cayce material.

Cayce's readings tell us of repeated natural catastrophes within the span of human existence on this planet, despite the fact that these events have escaped the notice of history. (Furthermore, the readings add greatly to the length of our existence on this planet; they indicate that we have been here for more than 10 million years.)

Cayce described Atlantis as a civilization that declined from a once virtuous and spiritual state into selfish materialism, producing internal strife and chaos paralleled only by the land's final physical destruction. Thus far, Cayce's account is quite similar to Plato's story. The major difference is that in Cayce's account, the Atlanteans possessed such a deep understanding of physical laws that they created, with relatively simple hardware, a sophisticated technology including flying machines and undersea craft, powered by a device called the "fire crystal" that somehow collected and focused solar and other cosmic energies for use on earth. The Atlanteans' use of energy was in harmony with their environment until the final days, when this energy was turned to destructive purposes.

In 1933, Edgar Cayce gave a reading on Atlantis that was apparently the first clue to Bimini's possible role in the puzzle: a remnant of an Atlantean temple might be discovered under the sea off Bimini Island in the Bahamas. Later, in 1940, Cayce predicted the rise of the western section of Atlantis: in twenty-eight years, or in 1968 and 1969. I am aware of no other source of information that drew attention to Bimini this early.

I was very skeptical, since material from psychics clashed with my conventional training in Western humanistic scholarship. But the following spring, I read Sheila Ostrander and Lynn Schroeder's *Psychic Discoveries Behind the Iron Curtain,* which led me to explore further the serious writings in the field. Various accounts of Russian parapsychological research, including their possible military applications, were beginning to appear in the West, and clearly showed me how much of our own published research was overly cautious, repetitive, even unimaginative. For example, no matter how many times the results of an experiment supported the phenomenon of telepathy, such phenomena could always be rationalized away. Why waste time and money to duplicate over and over what is transparently obvious to those who have investigated the field? Eventually I decided to investigate personally individuals who claimed various psychic abilities.

My first area to be scrutinized was the process of Kirlian photography. For several years a colleague in physics, Dr. Joseph Pizzo, and I conducted an experiment with this high-frequency process of photographing the "human aura." Our experiments convinced us of a strong correlation between a subject's physical, psychological, or psychic states and the Kirlian photos of his fingertips. (Details of these experiments are included in a book, coauthored with my wife, Joan, called *You Are the Mystery.*) Next, I found impressive evidence that some healing, both mental and physical, was possible by paranormal means. Equally intriguing was my discovery that many psychics could diagnose a patient's physical problem at a distance.

In all of this research, I began to develop a feel for what psychics could and could not do. I found, for example, that the weakest paranormal ability seemed to be prediction of the *dates* of future events. Although a psychic might correctly see the actual future event through clairvoyance, he could be all wrong on the timing. Taking my cue from Russian research, I sought to learn how psychic abilities functioned and what could be expected from them. As it developed, these years of university-based research into psychic phenomena opened to me many avenues of assistance that might otherwise have remained closed.

As I began to examine Plato's legend of Atlantis in the context of findings from astrophysics, geology, anthropology, archaeology, psychology, and parapsychology, I became more and more convinced that the tale could have had some basis in fact. Then in 1966, an archaeological development brought the Atlantis question into clearer focus: James W. Mavor, Jr., inventor of the research sub *Alvin* and a researcher at Woods Hole, claimed to have found Atlantis in the Aegean Sea. He compared the submarine profile of a burned-out volcano on Thera to Plato's account and thought he detected topographic similarities. Although I later came to doubt Mavor's claim that Atlantis was located in the Aegean, his serious approach reinforced my conviction that Atlantis was, indeed, a prehistoric possibility.

My own growing conviction that the Atlantic was a more likely site was bolstered by findings of a Swedish oceanographic expedition to the Sierra Leone rise. Science had finally begun to recognize that the Atlantic basin is unstable. Samples of the seabed suggested that perhaps 10,000 years ago the mid-Atlantic Ridge was above the surface of the Atlantic. These samples, or cores, found at a depth of 12,000 feet in the ocean bottom sediments, carried fossils of freshwater diatoms—a life form usually seen in freshwater lakes. Obviously, a land mass with fresh water existed here once.

In the summer of 1968, pilots Robert Brush and Trigg Adams, both ARE (Association for Research and Enlightenment, the Cayce organization) members who had sought tangible evidence of the Cayce predictions, saw what has since been called a "temple site" in shallow water just north of Andros Island, about 50 miles to the west of Nassau in the Bahamas and about 150 miles east of Bimini. The so-called temple is a stone foundationlike structure approximately 60 by 100 feet. The two pilots reported their discovery to Dr. J. Manson Valentine, a Miami archaeologist and zoologist, and Dimitri Rebikoff, who went to Andros and later described 3-foot-thick, skillfully worked limestone walls. Dr. Valentine recognized that the site's size and proportions also resembled the floor plan of the Mayan Temple of the Turtles at Uxmal, in the Yucatan. A news story dated August 23, 1968, released Dr. Valentine's account of the temple, which was located in 6 feet of water, its upper 2 feet rising *above* the ocean floor. Within a mile of the site, Valentine and Rebikoff found two other submerged structures. Since then, a total of twelve different underwater structures have allegedly been discovered in the Andros area.

On September 2 of the same year, while looking for various reported sites off Bimini, Dr. Valentine and others found two structures that they called walls, protruding about 3 feet above the sea bottom and extending for about 1,900 feet in a line which they thought paralleled the beach. This was half a mile off shore from Paradise Point, North Bimini Island, in the area since commonly referred to as the "Bimini Road." Dr. Valentine saw "an extensive pavement of rectangular and polygonal flat stones of varying size and thickness, obviously shaped and accurately aligned to form a convincingly artifactual pattern. These stones had evidently lain submerged over a long time, for the edges of the biggest ones had become rounded off, giving the blocks the domed appearance of giant loaves of bread or pillows. Some were absolutely rectangular, and some approaching perfect squares." He concluded that these huge stones were the remains of a megalithic site similar to that of Stonehenge.

In February of 1969 Valentine and Rebikoff joined forces with a Cayce group, MARS (the Marine Archaeological Research Society). During the last week of February, this group found another wall about 300 feet long and about 30 feet wide. Later, on March 15, Pino Turolla, another diver, brought up a rock from 15–35 feet of water on the Bimini Road site and noted that when struck, it gave out a metallic sound. (Our own research later revealed that the hardness of the limestone of these blocks resulted from chemical changes following the initial deposit of marine life and marine mud. This would later lead us to an important discovery about their origin.) Between July 12 and November 29 of the same year, Turolla found forty-four pillars 3–5 feet long and 2–3 feet in diameter, reported to be west of the Road site in 15 feet of water. An *Argosy* article described the pillars as from 3–6 feet in diameter and 3–14 feet long, with some standing upright, all located in a perfect circle.

In 1970 another survey of the features off Paradise Point was conducted by a group led by John Gifford, a graduate student in the University of Miami's Division of Marine Geology. Later he made two more surveys, one of them sponsored by the National Geographic Society and the University of Miami.

All of these expeditions were basically concerned with the same question: Are certain underwater features off Paradise Point the work of man or nature? Are they geological or archaeological in origin? So far, the findings have been ambiguous. Immediately after his second expedition, Gifford published a paper that affirmed that "None of the evidence . . . disproves human intervention." He later reversed himself and came out against the idea of human involvement in the site's construction.

As my studies of Atlantean theories continued, I became aware of a rather distressing phenomenon. The term coined by a biophysicist at the University of Pittsburgh is "anomaly anxiety." The conventional researcher trained in a specific area of expertise tends to experience this reaction when faced with new evidence that doesn't fit the prescribed parameters of his knowledge. Perhaps the best illustration is the (hopefully!) apocryphal story of the marine biologist who one day was handed a beautiful sea shell of a species new to him. Its very existence threatened to upset his lifetime of work on classification; now he would have to rethink this entire taxonomy. As he considered the specimen's implications, his distress mounted. Finally, he abruptly ground it to powder under his heel. "There," he said with relief, "it doesn't exist!"

To me, it became obvious that the specialist is more prone to anomaly anxiety than the generalist—and that the problem of Atlantis requires the generalist attitude. The Atlantologist must feel free to draw upon the work of many disciplines without concern about working in another's territory.

As one committed to this approach, in 1973 I took time to consider where it all was leading. The theoretical framework I was building contained many diverse elements: new ways of looking at ancient myths and the consciousness of prehistoric people; the earth sciences' new view of planet Earth as a complex mechanism involving the delicate balance of colossal forces; dynamic astrophysical relationships in the solar system; three years' investigation of cases of paranormal phenomena; Edgar Cayce's prediction about Bimini; the claims of Dr. J. Manson Valentine and others regarding Bimini. But my somewhat exotic theories were badly in need of support by some solid data, and I was not content with being a library scholar, as were so many others in this line of research. Ultimately, I decided to take leave from my university and undertake my own fieldwork to acquire firsthand data.

PART I

THE FIRST POSEIDIA EXPEDITIONS

1

GETTING UNDERWAY: POSEIDIA '74

I can still picture the faces of my fellow crew members that New Year's Day, 1974, as we motored out of Galveston harbor before hoisting sail. We were bound for St. Petersburg some 800 miles across the wintry Gulf of Mexico. As we watched the lowering clouds and passed the small-craft–warning flag at the Coast Guard station, the only cheerful one on board was my experienced mate, Eric. He and I had sailed this 9-ton sloop, *Makai II,* to Florida twice before; we knew she would make a fast passage, particularly as we were sailing on the tail of a norther. On this night, Eric and I enjoyed a marvelous run of 8-foot seas left from a gale that had almost blown itself out. But while we watched the phosphorescence of the churned-up foam, one of our two inexperienced young crew members was slumped in the cockpit, the other hung over the side seasick, both unimpressed by the beauty of the night.

On the third day, with one of our crew still sick below, a new gale came on us with sudden violence. Fortunately, the foregoing calm had made me suspicious and I had shortened sail. Within hours, the new wind made up 11-foot seas but we continued to hold our course and covered about 100 miles each day. To our relief, on the fifth day we broke free into the Eastern Gulf, which even in the winter is kinder to sailors.

About 100 miles off St. Petersburg we heard a radio report of possible sea fog. I sent a radar target aloft on a signal halyard to give us better visibility on steamship radars. Moments later, we heard the not-so-distant blast from a steamer's whistle. Although less than a mile away, it was invisible to us in the fog. Alarmed, I asked for our four-person rubber raft to be made ready. One crew member was already blowing the emergency signal on our own whistle. Seconds afterward, less than 500 yards off, a small British motorship came bearing down on us out of the fog. Fortunately her helmsman was looking sharp; he swung the 1,000-ton vessel away from us as gracefully as a small yacht. It took a few minutes for us to appreciate the narrowness of our escape from the near collision.

Once we had arrived at St. Petersburg, Ricky, our eighteen-year-old crew member, decided to continue on; the other two returned to Texas. My wife, Joan, daughter, Laurie, and son, David, joined us aboard. Ten days later, we had crossed Lake Okeechobee and were awaiting the right weather to sail from Ft. Lauderdale across the unpredictable Gulf Stream to Bimini. While the others kept an eye on the weather, I drove down to Miami for a meeting that would prove vital to our exploration of Bimini.

On the way down, I reflected on the events that had brought me here. The previous year I had realized that I had gone as far on library research about Atlantis as possible. In addition to my reading, I was counting on my experience in scuba diving and commercial photography to come in handy, too. I had also sailed the *Makai II* between Galveston and southern Florida several times, and considered her seaworthy for our mission. Unexpected assistance had come some months earlier when Charles Berlitz, in response to my suggestions for *Mysteries from Forgotten Worlds,* had introduced me to Dr. J. Manson Valentine, who now offered to provide me with some valuable background information on the Bimini area.

Upon my arrival, my host showed me many photographs of strange sea-grass patterns off the banks of Bimini, viewed over fifteen years of flights. Their shapes often suggested possible human-made structures beneath them. Dr. Valentine's material proved fascinating, and I was grateful for this introduction to the mysteries of the Bahamian waters. At the same time, I began to realize that a serious investigation of the Bahamas would be an enormous task, demanding scientific resources far beyond what my reading led me to imagine.

After our discussion with Dr. Valentine, we chose two sites for our initial five-week survey. One was the Road site near Paradise Point; the other was a triangular underwater depression that resembled a reservoir. Located just east of North Bimini Island, it was later labelled the "East Site." Armed with my new information, I was anxious to return to the *Makai* and begin our voyage through the Triangle—an apt beginning, I thought.

Back in Ft. Lauderdale we made the boat ready for a Gulf Stream crossing to Bimini Island. The center of the Gulf Stream has a 4-knot (nearly 5-mph) current that sets to the north, and winds with a northerly component that can quickly stir the sea into nasty going for any small craft. As we cast off from Pier 66, we expected a tough, wet night of it. Beating against the head seas, spray doused our foul-weather gear, and the hull often smashed down into hollows caused by the wind blowing against the current.

By 1 A.M. we had passed the worst. By 8 A.M., exhausted, we had tied up at Bimini to clear customs and immigration formalities. It was February of 1974.

The next day found us in a rented Boston whaler, running along incredibly white beaches off the island of North Bimini. Palms flailed their fronds in the trade winds and dark Australian pines stood still higher. It was our first trip to the Road site off Paradise Point, and we were all anxiously awaiting the first dive of the day. The bright Bahamas sun was overhead, the tide full, and the crystal-clear waters beautiful beyond description. To think that the Florida Keys had once enjoyed such water before overcrowding after World War II and people had clouded it. I tried to imagine these islands as Columbus first found them in 1492. At that time a simple culture of gentle Lucayan Indians (likely related to the peaceful Arawaks of the Caribbean) inhabited the islands. Their culture was one of thatched huts, baskets, pottery, and dugout canoes. Nothing in Lucayan culture suggested a connection with the huge stones off Paradise Point.

After Columbus discovered the Lucayans, the Spanish enslaved 40,000 of them and took them to Hispaniola to work in the mines, where many soon died. This left the islands uninhabited for nearly a century. During this time the present name for the islands evolved from the Spanish *bajamar* (shallow sea) into the present Bahama Islands, Islands of the Shallow Sea.

Now beneath these waters—such was their transparency—we saw glimpses of the huge blocks of Dr. Valentine's suspected megalithic site. Stunning in their size and obvious order, they could even be photographed from on deck through 18 feet of water.

Swimming with snorkels, some of us on compressed air, we were overawed by the cyclopean blocks of limestone. From a diver's perspective, they seemed to run endlessly. We had been in the water only moments when suddenly I spotted an 18-foot hammerhead shark moving slowly and majestically over the huge blocks below us. The hammerhead seemed intent on his own affairs, but I was afraid that if the others noticed him, they might become frightened and stir his interest by splashing about. I surfaced and in what I hoped was a quiet voice said "Everybody get in the boat," with no explanations. Once all were aboard, we leaned over the side with faceplates and I showed them the shark gliding along the bottom. We later learned that this big fellow was locally known as Harbor Master. Considering his size, the title seemed appropriate.

After our initial feeling of awe at the sight of the cyclopean blocks on the Road site, our diving off Paradise Point was rather frustrating for some days. The overall order of these blocks was unmistakable, particularly from the air, but once we closed in on them this order disappeared—at least temporarily. Like any subtle problem, this one required patience, persistence, and objectivity.

The Road site (which I ultimately came to regard as not a road at all) takes the form of a huge, reversed letter J. Its longer arm, about 1,900 feet in length, is composed of two parallel rows of megalithic blocks that terminate against a wider, pavementlike section of somewhat smaller blocks. The pavement section turns through a 90-degree arc toward the beach (a half mile distant). Extending the pavement a short distance brings you to two parallel rows about 327 feet long which make up the shorter arm of the J.

At length a clue to the order of the blocks emerged when we found what we have since called the "square stone"—about 8 by 10 feet in size—on the 1,900-foot lead.

When I first photographed this stone, it was clear that the block was surrounded by joints which rested against it on all sides, like a stone set into enormous masonry. For the newcomer to the site, it is probably the most compelling suggestion of human handiwork. Why this order existed was yet to be discovered; however, we noted the apparent artificial arrangement of these stones and went on to explore further.

From the first, our attention was drawn to the shorter leg of the J, particularly that area nearest the beach, because of the stones' great regularity. The beachward row also included one block, perhaps 12 feet square, one edge of which had somehow been lifted from the seabed and now rested against an adjacent block. We wondered at the fact that these blocks were not joined to the underlying bedrock. The obvious answer was that they were not originally shaped where they now rested. Not wishing to explore this any further at the moment, we

The main lead on the Road site.

contented ourselves with photographing the blocks, emphasizing the joints.

Several days later, we turned our attention to the East Site. Perhaps as large as 100 acres, its southern edge is bordered by what appears to be a dike, with parallel rows of rectangular shapes extending several hundred yards. Our aerial photographs show hydraulic action from tidal currents diverted by this dike. The shape, in reality, is formed by turtle grass, which could have covered stones that once lay below and have since dissolved. Our speculation was that this site may have been an ancient reservoir.

Our curiosity whetted, my wife, Joan, and I decided to fly to Nassau for an aerial view of the temple at Andros Island reported by Trigg Adams and Robert Brush in 1968. We could see it clearly during the Chalk's Airline flight, and its distance from the shore seemed to argue against its recent construction. Yet just the year before, John Keasler, a reporter for the Miami *News,* had interviewed fifty-five-year-old Reuben Russell, constable on Andros Island, who claimed he had helped build the structure in the 1930s for a Nassau man who used it as a holding pen for sponges awaiting shipment. Dr. Valentine and others found eleven more geometric shapes around the northern waters of Andros Island including a hexagon, parallel lines, and a semicircle on one end of a straight line. Also visible is a huge reversed E. Since these represent a variety of patterns, it does not seem reasonable to claim that they were all sponge pens.

Just before finishing our five weeks of research in the Bahamas, we took a fathometer profile of the two shorter rows of blocks at Paradise Point to determine the plane of the seabed. Our findings showed that the bottom was parallel to the water's surface, not sloping. Since beach rock is usually formed on a slope, this seemed to argue against the blocks being beach rock, or naturally formed on the site.

The upshot of our work in 1974 was a growing feeling among us that the so-called Road site was not a natural formation but could, instead, have been the result of human engineering. The straightness of its 1,900-foot length suggested the precision of a survey. The order of the various rows of blocks; the fact that they were not attached to the seafloor; our chartwork that showed their overall orientation about 7 degrees out of alignment from the present beach line—all led us to conclude that we were likely dealing with an underwater archaeological site. If Dr. Valentine's theory of Bimini as a megalithic site were valid, perhaps it would be a good idea to find a specialist in this field to join us in the search.

Once a suspected temple site, this structure is now believed to be a sponge pen built in the 1930s.

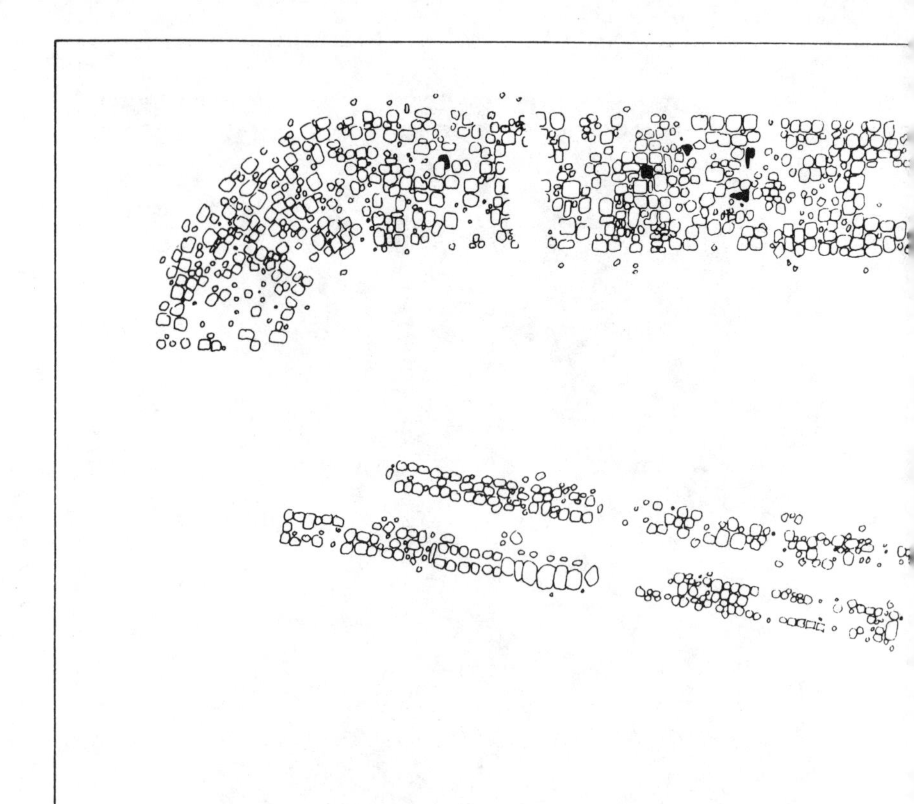

The J-Formation of the Bimini Road.
(Courtesy John Parks.)

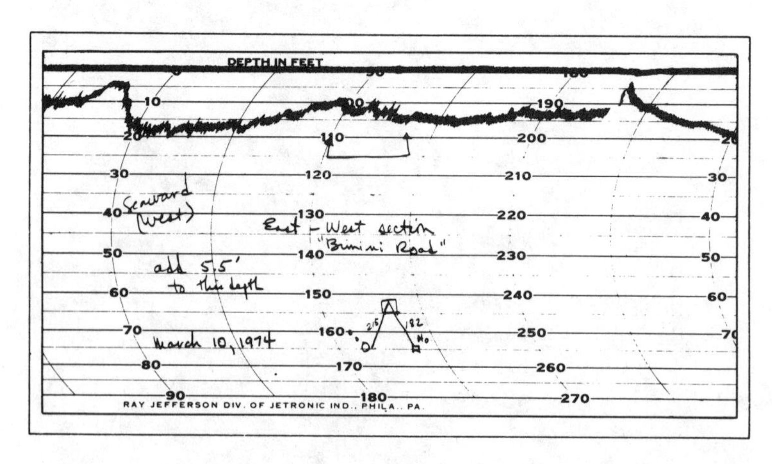

Left: *Fathometer profile of the Road site, which is half a mile offshore and* not *on a slope as is usually the case with beach rock formed in place.*

The J-*Formation.*

2

BIMINI AS A MEGALITHIC SITE

On the eve of the 1975 expedition to Bimini, we were sitting in Le Joint, a popular waterfront restaurant in Miami, listening to archaeologist John Steele—in his early thirties, dressed in English bohemian attire—describe megalithic sites in England and Ireland. A veteran of Dr. Maxine Asher's 1973 underwater expedition seeking evidence of Atlantis off the coast of Spain, John had flown over from a lecturer's post at University College in London. His background also included involvement in scientific projects funded by NASA and the National Science Foundation; and he had just presented a paper with John Michell, the British expert on megalithic sites, to a symposium on ancient energy.

Menhir at Stonehenge.

John, who had yet to visit Bimini, suddenly turned and asked me: "Do you really think it's a megalithic site, as Valentine has claimed?"

"Seems that way. What we've seen so far favors the man-made theory."

"What's your theory about the function of the Bimini Road site?"

I explained that I didn't think it was a road but, due to its shape, more likely related to some ancient harbor works.

John felt that whatever its function, the mere presence of such large worked stones certainly didn't fit in with any known culture pattern of the Bahamas.

Two dolmens.

He proceeded to review the basic features of megalithic sites. Megaliths, or huge stones, are elements of prehistoric monuments and architecture. Each stone (or monolith) is found either standing as a menhir (single upright stone), in circular patterns around a mound, or supporting a capstone, a dolmen. They may be either natural or partly carved or dressed. Because bones, tools, and apparent offerings have been found in some of the dolmens, it is felt that they (and perhaps all megalithic sites) were once tombs, at least in part.

John pointed out that one of the intriguing aspects of

megalithic sites is the size and weight of both their natural and arranged stones. The Egyptians, for example, used 70-ton granite blocks in the King's Chamber of the Great Pyramid. At Stonehenge, one of the huge standing stones within the circle has a total height, above and below ground, of 29 feet, 6 inches and weighs about 50 tons. In France the great menhir at Locmariaquer, once erect as a monolith and now in four pieces on the ground, measures 77 feet and weighs more than 300 tons. The controversial Mystery Hill megalithic site in New Hampshire includes stones weighing more than 6 tons. The larger blocks at Bimini are probably about 15 tons. The sheer size and weight of these stones raises many difficult questions. Sometimes they are superbly carved and shaped, often a part of superior masonry construction, which compounds the puzzle of their origin. The explanations that they were fashioned by stone implements or hardened bronze tools are convincing enough until you actually inspect them closely. Their intricacy of design seems far beyond what rudimentary tools could accomplish.

I realized that John, like myself, used the term *megalithic* in a broader sense, to include New World sites like Tiahuanaco in Bolivia and Machu Picchu in Peru. Some of the carved stones at Tiahuanaco, for example, weigh on the order of several hundred tons. The rougher monolithic stones in the main temple are as tall as 9½ feet, approaching the nearly 14-foot height of Stonehenge's great trilithons. The most important thing about megalithic culture is our recent awareness that their amazing structures are based on a much more evolved science than either Neolithic or Bronze Age people were ever thought to possess. Built into their architecture are sophisticated astronomical alignments, such as solar and lunar risings and settings.

Sarsen circle at Stonehenge.

From the ground level, Stonehenge is the most impressive of the European megalithic sites. The more that is learned about it, the less is understood of its origins, functions, or the unaccountable plan of its master builders. Whoever constructed Stonehenge had a surprising awareness of aesthetics; the lintelled sarsen (sandstone) circle which the visitor first sees from the road gives evidence of an architectural technique common in the columns of classical Greek temples. To correct the effects of perspective, the sarsen stones, which taper as they rise, are also curved convexly. How did this level of cultural awareness find its way to England's Salisbury Plain during the Stone Age?

Gerald Hawkins, who successfully attacked the problem of Stonehenge with modern computers, worked from a photogrammetric survey to establish the possible sighting lines of astronomical phenomena with greater accuracy than anyone had before. He fed this and astronomical data into a computer and discovered that Stonehenge provided sighting arrangements for the sun at the times of the equinoxes and solstices and also for the extremes of the moon. He made the further startling discovery that the site could be used as a computer to predict lunar eclipses. Since this time still more accurate surveys directed by Alexander Thom have reinforced the lunar eclipse prediction idea.

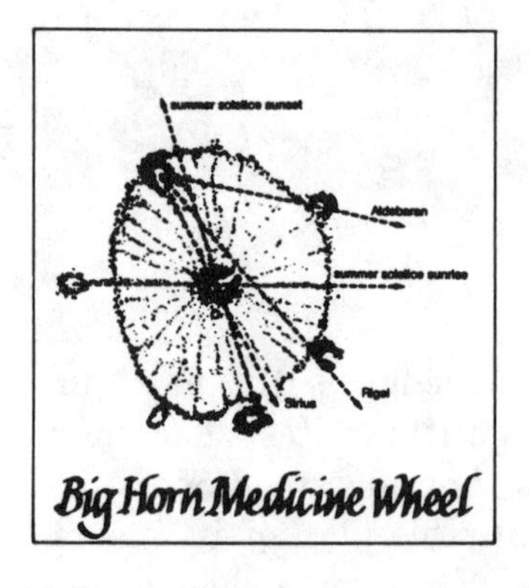

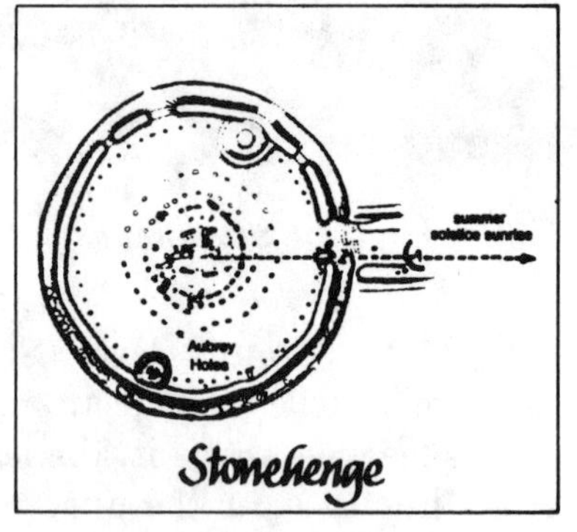

Note the common orientation of both megalithic sites toward the sunrise. (British Crown Copyright. Reprinted by permission of the Controller of Her Brittanic Majesty's Stationary Office.)

By 1974, the year of my first expedition to Bimini, the new discipline of archaeoastronomy had established itself to the extent that an astronomer, writing in *Science,* could say: "The solstitial alignments of Stonehenge and other European megalithic monuments, of

Egyptian pyramids and temples, and of Mayan temples are by now generally recognized." The writer John A. Eddy, at work on an American "Stonehenge"—the Medicine Wheel in the Big Horn mountains of northern Wyoming—had moved off to more sophisticated astronomical phenomena, the heliacal risings of stars. (A heliacal rising, important to the ancients for calendrical purposes, occurs when a star rises momentarily before sunrise and is then extinguished by the light of the predawn sky.) The Egyptians used the heliacal rising of Sirius to help time the summer solstice, which signalled the annual rise of the Nile. The annual flooding of the fields by the Nile was necessary for productive agriculture, but the people required warning to take to the safety of the hills. Eddy's investigations led him to claim that in Wyoming the summer solstitial observation was accompanied by the heliacal rising of Aldebaran, the bright star in Taurus. Obviously, solar and lunar risings and settings, the meridian transits of stars, the heliacal risings of stars, and even predictions of eclipses were important observations for ancient astronomers, who incorporated these phenomena into their megalithic structures.

STONE AGE COMPUTERS

Because there is so little evidence about megalithic culture, every new site presents exciting possibilities, particularly when it is found on another continent. New Hampshire's Mystery Hill, once believed to have been built by a nineteenth-century farmer, includes stones of over 6 tons that mark the summer and winter solstices. Dr. Barry Fell, professor of invertebrate zoology at Harvard University and president of the Epigraphic Society, (a group that examines and translates ancient stones), has translated inscriptions to the Celtic sun god Belos, and Baal, the chief deity of the Phoenicians, from a single stone at Mystery Hill. This evidence strongly suggests the presence of two ancient cultures in America at least as far back as 800 B.C. The earliest occupancy of the site, based on a corrected carbon-14 date, may be as early as 2000 B.C. At present, the group exploring the site—the New England Antiquities Research Association—is at work on another site in Vermont that also allegedly abounds in Celtic artifacts and inscriptions.

Megalithic structures share far more subtle features than large, heavy stones do. However, John Michell—today's most serious writer on the subject—has shown that prehistoric peoples felt themselves, their temples, and cosmos to be parts of one life system. All true megalithic constructions, we learned, are based on "sacred engineering." Thus, the code of a civilization, its spiritual awareness, and technological harmony with the environment were actually built into the structure—not plastered on later. The buildings contained measurements that are now being decoded by gematria, the system of correspondences between numbers and letters. Also discovered within these structures are geometrical and geodetic proportions such as pi, phi, and equivalents of the earth's rotation and the orientation of the site in relation to the stars, sun, and moon, other megalithic sites, important mountains, islands, and so on. The presence of such elements reflects the idea that the cosmic temple was a living organism with the capacity to resonate with all other living systems. This is a truly awesome idea whose implications take some time to grasp.

"Megalithic constructions," John continued, "invariably served a sacred function in the ancient world, the

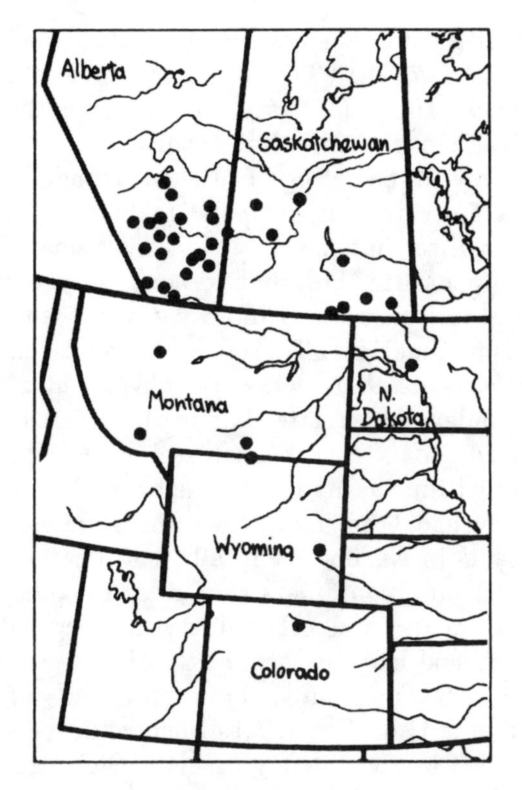

The locations of wheel sites in North America. (British Crown Copyright. Reprinted by permission of the Controller of Her Britannic Majesty's Stationary Office.)

purpose of which was to integrate man with, rather than to isolate him from, his environment, to connect him simultaneously to the macrocosm of the stars and the microcosm of the atom." This view is based on the hermetic tradition, "as above, so below." In order to reconcile elements of the earth with the heavens, one needed the equivalent of a transformer, such as that which allows the output of an audio amplifier to be matched with a speaker system. As John Michell said in his *City of Revelation:* "The cosmic temple was an equilibrator whose function was to reconcile all the diverse and contradictory elements of nature."

During my first reading of Michell, I understood this in only a figurative or symbolic fashion. Gradually I began to realize that not only does mathematical order originate in nature, but that in some as yet unknown fashion, geometric shapes intensify and focus various subtle cosmic energies. Therefore, I was fascinated by *Psychic Discoveries Behind the Iron Curtain* by Schroeder and Ostrander, who relate that in the 1950s a Frenchman named Bovis found small animal carcasses in a waste can within the King's Chamber of the Great Pyramid. They had been mummified and gave off no odor. Experiments in Europe eventually led to patented pyramid-shaped milk containers that retard spoilage, razor blade sharpeners, and other devices. I learned that a physicist friend had conducted extensive experiments in Houston, Texas, and found that even in the warm moist climate of the Gulf Coast, fresh fish could be similarly mummified. Clearly, the dry desert air of Giza was not the only factor at work.

Many scientists have offered plausible explanations of techniques for moving the gigantic stones at Stonehenge and Easter Island. And, perhaps with a large, well-organized society such as existed in Egypt, huge stones with weights up to 1,000 tons (in the case of the pink granite obelisks brought many miles down the Nile) could be managed with the suggested techniques. But is it conceivable that a simple people could have used these more sophisticated methods? Even in the Egyptian culture, we cannot account for the incredible precision with which many of the stones of the Great Pyramid (c. 2600 B.C.) are arranged. In the 1880s one investigator, applying precision instruments to some of the 15-ton casing stones he had unearthed, found that the average variation both from a straight line and a perfect square was only one one-hundredths of an inch in 75. This is the accuracy of skilled optics, not crude masonry. Work of this quality must be troublesome to those who still assert the ignorance of the Egyptians. Writing in the 1960s, one particular commentator, described by Peter Tompkins in his book *Secrets of the Great Pyramid,* made the outrageous statement that the Egyptians were unaware of the cardinal points of the compass except for east and west—which they knew only from sunrises and sunsets. Could such a culture have accomplished the work of building the Great Pyramid?

Later, in experiments of my own, I discovered that a model pyramid large enough to contain a person and properly oriented to true north seemed to enhance the subject's alpha brainwave activity and also stimulated several involuntary out-of-body experiences. The latter discovery was really no surprise because esoteric sources claim that this was one function of the Great Pyramid when it was used for initiation purposes.

In various attempts to probe the nature of the energy generated by a pyramid in relation to its orientation, I blindfolded a sensitive woman and had her "tune" the pyramid for maximum energy by rotating it. When the pyramid's side was aligned with true north, the psychic reported the maximum energy. The energy output continued noticeably when the pyramid was within 10 degrees (on either side) of true north. Other blindfolded sensitives reported maximum energy above the apex of the pyramid. All of these clues should have helped me to understand Michell's term "equilibrator" as a scientific quantity, whether or not we presently have the science to understand it.

In my research I had discovered that early humans were apparently capable of recognizing strong currents of magnetic energy at various points on the earth's surface. From China to Britain, ancient architects employed sensitives known as geomancers to help locate their sacred structures at high magnetic energy points, where human activities would be in harmony with terrestrial currents. These power points, marked by standing stones (and later, often by churches), were connected by invisible lines of magnetic energy called "ley lines."

Today a more comprehensive view of these energy patterns is emerging, and they appear to be global in extent. Writing in the *New Age Journal* (no. 5, 1975), Chris Bird, coauthor with Peter Tompkins of *The Secret Life of Plants,* suggests that some scientists are now seriously entertaining the ancient notion of metaphysics that "our Earth and all matter upon it, whether held to be 'living' or 'non-living', is but the final result of a transformation of energy. . . ." His article, "Planetary Grid," summarizes several Russian theories about the existence of a global energy pattern. The earth is seen as a crystalline structure whose energies are reflected up through power points on a dual system of geometric figures (twelve pentagons and twenty equilateral triangles) that cover the earth and have significant intersections in the Bermuda Triangle (which includes Bimini), at Giza in Egypt, and in Peru. These and other intersections also happen to be maximal points of solar radiation and the centers of the unexplained magnetic inconsistencies on the earth. This planetary grid system's lines appear to coincide with weather patterns (including hurricanes), midoceanic ridges, and the edges of continental plates.

For example, as one vessel crossed the Puerto Rican

Trench during a Caribbean voyage, a sensitive independently identified a magnetic anomaly identified by instruments aboard. During my own investigations, one of the sensitives, a graduate student at my university, told me that her sensitivity was dramatically enhanced whenever she visited her mother's home in west Texas. At length, consulting a magnetic survey map, she found that the spot where her mother's home was located was characterized by strong magnetic anomalies.

John Michell's excellent study, *City of Revelation,* enumerates several components of megalithic sites, among them: (1) an astronomical orientation and a location in relation to other sites and terrestrial indicators, such as mountains and magnetic fields; (2) numerological features expressed in the dimensions and layout; (3) systems of geometry corresponding to significant numbers, reflected in the ground plan of the temple.

The first criterion is covered (at Stonehenge, for example) by a site's lunar and solar sighting arrangements, its situation on lines of magnetic force within the earth, and its relation to underground water courses; (the latter two, I suspect, are also connected to the cosmic energies generated at sacred sites.) The second criterion is satisfied when the site includes the number six, (related to inorganic structure such as crystals); the number five, (related to organic life); the "spiritual and esoteric" number seven, sacred in many cultures; and the twelve of the lunar cycle. For the final criterion, geometric figures related to these numbers, such as the circle, square, cross, triangle, hexagon, pentagon, and others may be found in the ground plan of the temple. (For more information on sacred geometry, see appendix D.)

How well, I wondered, would Bimini satisfy the crite-

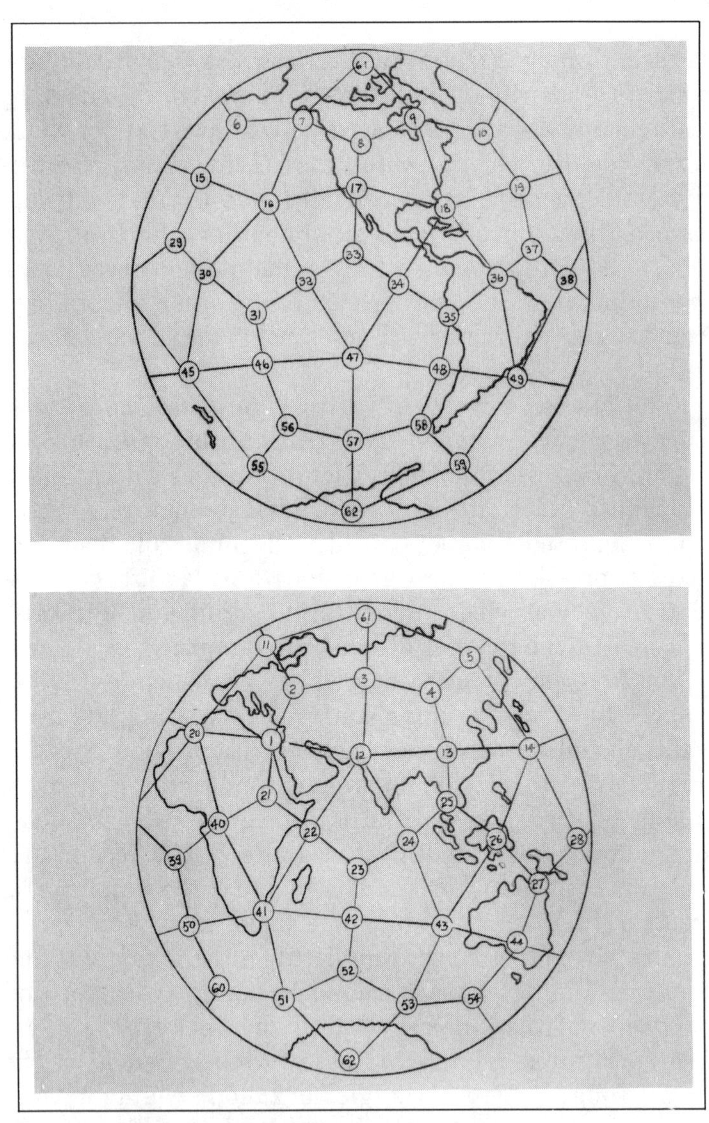

The Russian planetary grid system.

ria for a megalithic site? The enormous blocks were one indicator, but what of the sacred geometry, the unusual magnetic activity, and the astronomical orientation? Would we be able to find such indicators? John Steele's highlights of megalithic sites had given us much to ponder and, perhaps, to discover.

3

NEW CLUES TO THE PAST: POSEIDIA '75

Once again we had arrived at Bimini after a Gulf Stream voyage which, happily, proved milder than our first in '74. John Steele and C. W. Conn, my diving partners, were as expectant as I. For this first day on the Road site for the Poseidia '75 crew, the weather looked good—no towering thunderheads or threatening squall lines were to be seen in the distance, and the Gulf Stream's giant weather machine was quiet. This time we were aboard a 60-foot dive boat, *Fosi III,* and C. W. and I were readying our dive gear for a preliminary look at the site. Even at 9 A.M. the tropical sun was bearing down on us, but we were a bit early for the best light. Despite the clarity of the water, a high angle of the sun was still desirable.

Diving from the *Fosi,* we descended to the megalithic blocks below, where I got a bearing on our location. We were now over the long outer row of blocks, near the center of the long arm of the J. Swimming in about 15 feet of water, we found ourselves surrounded by a colorful assortment of tropical fish. For them, the site was a reef sheltering them from larger fish, but they seemed unafraid of us as long as we kept our distance.

While investigating the blocks' positions, I suddenly noticed, beneath the jointing of two of them, a fracture in the solid limestone seabed. What was significant was that the fracture did not coincide with, or even parallel, the joint above it. In fact, the two lines of joint and fracture intersected at about 45 degrees. I beckoned C. W. over to have a look.

Taking measurements, we found that the joint between the two blocks was oriented 315 degrees toward the northwest, at a right angle to the orientation of the row of stones themselves, which were 45 degrees northeast. Checking the seabed fracture, clearly the work of nature, we found it generally oriented toward 0 degrees azimuth, or north. This certainly raised some questions! Earlier, geologists attempting to explain the jointing pattern of these huge blocks had claimed that natural stresses, or tensional jointing processes, were responsible. If this were so, the joint between the blocks should coincide with the seabed fracture. We had found a possible contradiction of this theory on our first dive. During the expedition, we would later find two more identical cases where natural fractures in the seabed bore no significant relationship to the joints of the megalithic blocks.

Swimming over another part of the site, we became aware of another phenomenon. A row of four or five blocks seemed to overlap the blocks beneath them; they appeared to have tumbled over and seemed situated in a most unnatural way. Earlier surveys had reported *no* stones on top of other stones.

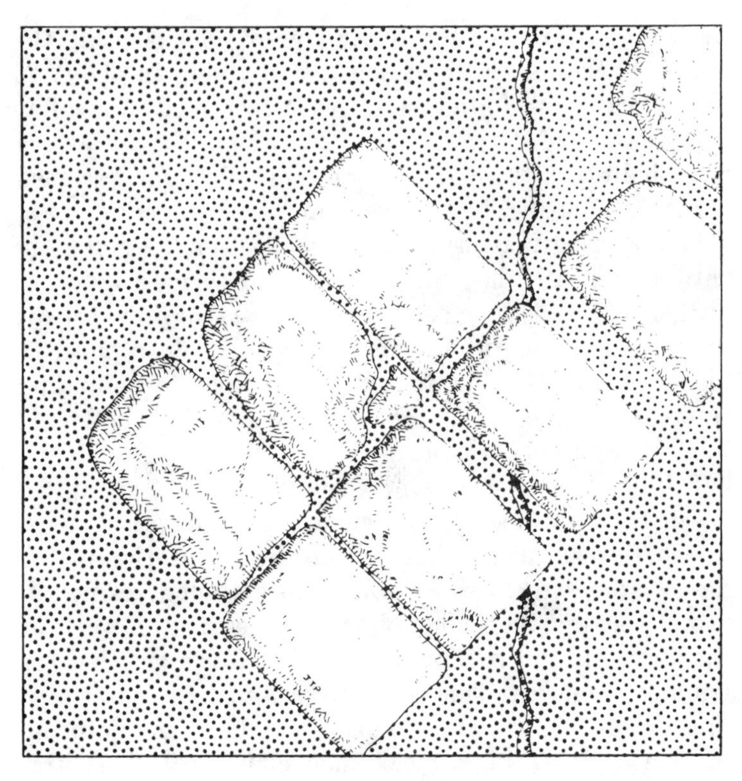

Blocks jointing in apparent contradiction to natural stress in the seabed. (Courtesy John Parks.)

That evening after dinner we discussed our findings, then turned to the more practical aspects of our coming weeks on the site. Duties were assigned and health suggestions offered, as well as the need for courtesy and consideration of others in the close quarters and debilitating tropical climate. I advised using salt tablets to prevent dehydration, and alcohol in the ears after diving to avoid inner ear problems. Everyone was reminded of the Bahamian law against the use of compressed air or scuba gear to gather lobsters or hunt fish; only snorkeling was permitted while fishing. We also stressed the

The wedge-shaped "chockstone."

need for protection from the tropical sun. Then everyone signed the articles of agreement of Poseidia '75. Members of the expedition were quartered aboard the *Makai II* and another sloop, *Gypsy,* kindly loaned to us by Frank Spampinato, a graduate geologist.

Our second day was as productive as the first. Farther south on the main lead, and within a few feet of the square stone, we found what I called the "chockstone." Its shape, like a section of pie with the apex cut off, reminded me of stones blocking natural chimneys that I had found on mountain climbs. It also resembled the keystone of an arch. Its central axis ran north–south (magnetic). Looking south, each side opened 30 degrees. This pattern seemed far too orderly to have had a natural origin, and as we studied the blocks on the days that followed, the chockstone soon emerged as a significant part of a geometric pattern extending about 60 feet to the east and about 60 feet up the row of stones to the northeast. I wondered whether this might be a clue to a sacred geometry as described by Michell.

One of our most important objectives was to accomplish the first full tape-and-compass survey of the site, ultimately producing a map of the site based on aerial photography as well. We were fortunate in our timing because the continuing shifting of sand had worked in our favor. Except for one section, much more of the site was visible to us than the year before.

Most of the blocks were now clearly resting on either the underlying bedrock or on smaller stones on the seafloor. Besides assisting us in our survey, this fact had an important archaeological consequence: it meant that the view (held by some Atlantologists) that the blocks now visible were only the top of a more complex structure was likely incorrect. Although it's still possible that lower layers of the structure may have been cemented under the marine limestone of the seafloor, the present site itself is clearly limited to a depth of one, and sometimes two layers of stones.

It was becoming obvious to us that we had to have a greater familiarity with the geology of the site and the island, including the processes by which limestone, the principal sedimentary rock, was formed. We also needed to learn about the exact composition of other types of limestone in order to eliminate the immediate area as the source of the giant stones. Existing geological surveys of Bimini did not answer these questions to our satisfaction.

The arrival of John Parks brought us the technical expertise we would need to review the geological problems. A young geologist, John was an instructor at a university in eastern Tennessee, the youngest member

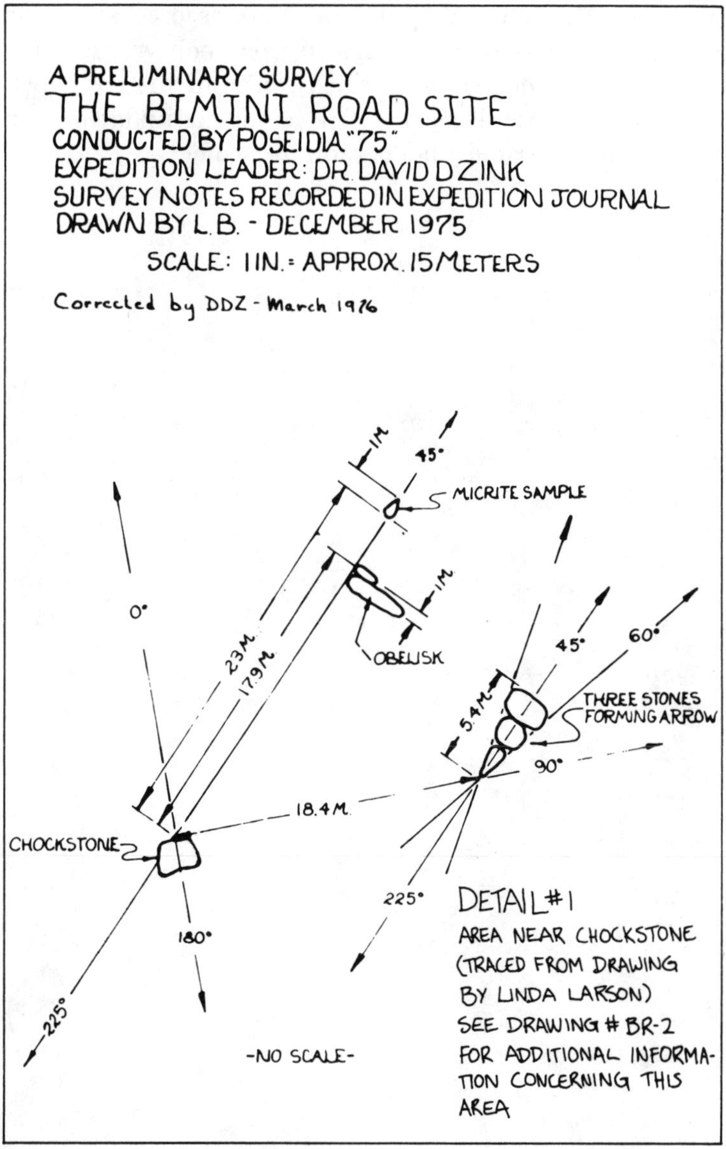

A PRELIMINARY SURVEY
THE BIMINI ROAD SITE
CONDUCTED BY POSEIDIA "75"
EXPEDITION LEADER: DR. DAVID D ZINK
SURVEY NOTES RECORDED IN EXPEDITION JOURNAL
DRAWN BY L. B. - DECEMBER 1975
SCALE: 1 IN. = APPROX. 15 METERS
Corrected by DDZ - March 1976

1M
45°
MICRITE SAMPLE
1M
0°
23M.
17.9M.
OBELISK
45°
60°
5.4M.
THREE STONES FORMING ARROW
90°
18.4M.
CHOCKSTONE
225°
180°
225°
-NO SCALE-

DETAIL #1
AREA NEAR CHOCKSTONE
(TRACED FROM DRAWING
BY LINDA LARSON)
SEE DRAWING # BR-2
FOR ADDITIONAL INFORMATION CONCERNING THIS
AREA

The Bimini Road site. (Courtesy Linda Larson.)

of his faculty and, to my delight, possessed considerable artistic skill (as his drawings in this book will reveal). In the water, he was a certified scuba diver; on board, John was particularly helpful to me in updating my twenty-five-year-old undergraduate geology.

4

BAHAMIAN GEOLOGY: THE STONES BEGIN TO SPEAK

In our research we were investigating the makeup of the main unit of the Bahamas, the Great Bahama Bank with the Bimini group in its northwestern corner, and Andros Island on its eastern edge. To the north, over the deep North West Providence channel, is found the Little Bahama Bank. The third bank, to the west of the other two, is Cay Sal Bank, mostly submerged. The entire archipelago includes seven hundred islands in an area of about 40,000 square miles.

As a geological unit, the Bahamas date back at least to the Cretaceous period. Their age for the deeper limestones is about 130 million years; exposed strata are much younger. During this vast time, the underlying bedrock has slowly been subsiding, and carbonate rock has gradually built up on the subsiding platform—a plate called the Bahama Platform, related to the continental plates of North America. As the plate has gone down, reef-building organisms (including coral) have built up sedimentary deposits of carbonate rock at least as fast as the plate has sunk. The result is that for many thousand feet beneath their surface, the Bahamas are composed of carbonate rock or various types of limestone.

The bedrock of the Bimini islands is eolianite, a windborne limestone formed during the Pleistocene period that has been named Old Bimini. Under microscopic analysis it is seen to consist of medium-size grains (oolites) cemented together in a matrix of calcium carbonate that takes the form of blocky crystals (sparry calcite). The oolites themselves are nuclei of calcium surrounded by concentric layers (or radially oriented needles, like a porcupine's) of calcium carbonate. Essentially these grains were sorted by wind action, then cemented by the crystallization (from a supersaturated solution of seawater) or blocky structures of the mineral form of calcium carbonate. The technical name for this windborne limestone is intraooparite, or oolites cemented by sparry calcite. The alternative cement is aragonite (long needlelike crystals, yet another form of calcium carbonate).

The layer above the Old Bimini bedrock is known as New Bimini. This limestone is coarser, and much more abundant in fossils. The cement of these grains touches only the grains themselves and does not fill the porous space around them. The result is a much softer rock. It has been dated to the Holocene age (within the last 10,000 years), and may have been laid down as storm deposits in this most recent of the geological periods.

The third major type of rock native to the Bimini group is beach rock, some of which has been formed quite recently. It is located in the tidal zone, and is thus sometimes uncovered. While the exact process of its formation is not as yet clear, one thing is certain: under the right conditions it can be formed quite rapidly, as evidenced by beer bottles found cemented within. At present beach rock is being formed of boulders of New Bimini limestone set in a cemented sand; *Strombus* conch shells, coral skeletons, limestone pebbles, carbonate crust pebbles, and pieces of glass bottles. These elements are all being cemented together by aragonite crystals oriented radially around each grain.

The fourth major rock type of importance to our investigation is a marine limestone that, even to the untrained eye, varies considerably from the megalithic blocks. Its grains are much finer and well sorted, with no evidence of the larger grains found in the blocks themselves. It is a biopelsparite—pellets (or grains) of biological origin cemented by sparry calcite.

John Parks immediately busied himself collecting hand samples of the blocks so that later he could prepare thin sections for microscopic analysis. He also probed some of the nearly buried joints between the blocks, using a floating water pump which, though small, produced 100 pounds per square inch at the nozzle. Separating the blocks at one point, he found a narrower joint than between those blocks that had been more exposed to various kinds of erosion. In fact, the joints were so narrow they resembled masonry. We would eventually spend many patient hours underwater in a study of the arrangement and configuration of the blocks to see if their structures had more in common with natural formations or with ordered human construction.

Studying the blocks underwater to determine whether their arrangement is planned or accidental.

Late one afternoon, I sought out John Parks and finally found him busily arranging some of his samples taken underwater off Paradise Point. As we sat on the edge of the dock, I finally asked a question that had begun to assume a great deal of importance over the last few weeks: did he feel the rocks we had been investigating were beach rock, formed in place, or rock of another material brought in by early peoples? I was surprised that John had come to the conclusion that they were most likely beach rock. However, he suspected that the Road's arrangement was too orderly to be naturally formed, and had confirmed that these particular rocks' rectangular shapes were unlike those found elsewhere in the area. The local beach rock was much softer and far more disordered in its patterns.

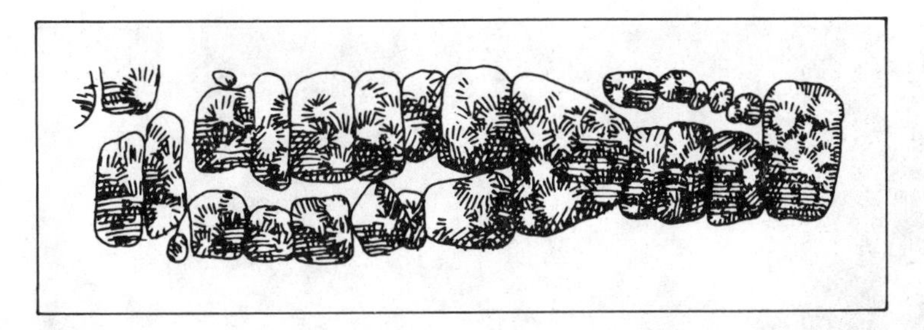

"Five" and "six" patterns in the stones. (Courtesy John Parks.)

Perhaps the most outstanding structural feature of beach rock is the development of equally spaced ridges and furrows that are nearly parallel. These furrows form at a right angle to the beach line, perhaps through erosion by sand carried in the surf and tide. The best-known examples of beach rock contain an average spacing between furrows of about 24 inches and a furrow width of about 11 inches. Our initial visual impression of the variance between the patterns of the beach rock and the megalithic blocks began to be accounted for by our later measurements. What would correspond to the furrow spacing, the length of the blocks, is on the order of about 10–13 feet. The 11-inch furrow width of beach rock does not at all match the spacing joints between the megalithic blocks. Here we found two distinct patterns: one which ranged from about 27 inches to about 31 inches, and the other from about 4–6 inches.

Scientifically speaking, this is one aspect of the morphological side of the argument for a human hand on the site that had not been given sufficient attention before our expedition. Because of the sublety of the lithology of the blocks (the result of the microscopic examination of their components), morphology becomes even more important than usual.

Diver atop the "square stone," which is resting on smaller stones.

But even more convincing to John was the aerial view of the site, particularly the area where the pattern changed from two sets of parallel rows of blocks to the pavementlike section. He also regarded the 90-degree arc in the pavement's turning as evidence of human design.

John felt, however, that the strongest evidence supporting human intervention lay in the pattern of joints between the rocks. In one area five large stones lie in a line almost parallel to the beach, with five smaller stones just seaward of them. South of the five large stones, the pattern continues with six stones equal in size, but their long axes run parallel to the beach. Thus, the longer joints in one set of blocks are at right angles to the lon-

The "square stone": smaller blocks supporting the larger blocks refute earlier assertions that the positioning of the blocks is consistent with the formation of natural beachrock. (Courtesy John Parks.)

ger joints of the other. This symmetry and apparent design within the short distance of only 40 feet defy any natural explanation.

More evidence supporting John's theory began to appear when we discovered that in the seaward lead, one of the largest stones' corners was resting on smaller blocks. One day, when looking at the square stone in clear water, we were amazed to see that its corners too rested upon smaller blocks, as did one adjacent block. This formation recalls similar arrangements at other megalithic sites, and are suspected by some of having been able to boost magnetic energy. These findings challenge one of the pieces of evidence cited against the man-made theory by John Gifford, a geology graduate student at the University of Miami. After investigating the Road site, he stated that no evidence exists anywhere among the three groups of blocks of *two courses* of blocks, or even a *single* block set squarely atop another.

We now sought to analyze samples of the rocks to determine their composition. Later, when John Parks visited me in Virginia, we visited Old Dominion University to learn the results of the analysis. They were intriguing, to say the least: it seemed that the cement of the various sections—composed of marine life forms and crystalline forms of calcium carbonate—was not alike. One sample was dominated by aragonite crystals, another by sparry calcite. This implied that adjacent stones were formed in different chemical environments. To be sure I decided to take a further step: to drill into the centers of about a dozen of the blocks to determine whether the layers of deposits matched in blocks resting beside each other. Although an expensive operation to consider, it was the only way we would be able to settle the question of whether the blocks were formed on the site. Where, I asked myself, would the money come from to finance this undertaking? But any concerns about money needed for the coring—slated several months hence—began to seem less important as we found ourselves face-to-face with the day-to-day problems of island living.

5

BUGS AND BARRACUDA

When Poseidia '75 members first settled into the three-room flat donated to us, we had been delighted with all the space. But as new members of the expedition joined us, the number grew to over a dozen, and conditions became less and less comfortable. Apart from the problem of extra people, the toilet did not flush without its tank being filled with a hose passed through the bathroom window. The tub drained its used water at a rate of an inch an hour, the water accumulating unpleasantly when one shower followed closely on another. Our venerated air-conditioning unit, although used only at night, fell victim to the vagaries of the unpredictable island power plant, and those sleeping in the flat had to resort to leaving the windows open rather than suffocate. Before long they became easy targets for myriad biting bugs and mosquitoes, and those who continued to sleep in the flat were often exhausted and wrapped mummylike in their sheets the next morning. It usually took a messenger from the boat to rouse them.

Evenings could be equally taxing. One wet, humid night in particular, we had set the tables together in the living room to accommodate a larger group. Since some of us had become so overheated that we opened the sliding glass doors from the porch, the mosquitoes were particularly bad. Bravely we sat down to eat amid the humidity and bugs, many having already sprayed themselves with insect repellent. Soon smoke began to waft through the open windows. Outside, the edge of the island graveyard was afire; despite the rainy evening, fires from the dump still further south had gotten out of bounds. This seemed inexplicable but hardly more incongruous than any dozen daily occurrences. All of a sudden the preposterousness of our situation hit Joan. In half-humorous despair she said, "What are we all doing here?" In our own separate ways, each of us was tested during the summer, sometimes to the edge of endurance.

When the *Makai II* had sailed from Galveston to begin the expedition, she carried 1,500 pounds of gear—including a comprehensive inventory of spares for all likely contingencies. But soon I had to rebuild *Makai's* head. Then her galley fan burned up. Items of rigging eventually had to be improvised for both *Makai* and *Gypsy.* The breakdowns continued, and now my concern began to grow as I watched a generous store of spares disappear with each new breakdown.

One day, just after lighting off the engine of the *Makai,* I heard a terrible clatter from the propellor shaft. I immediately realized that we had serious problems—either with the shaft, its bearings, or both. The hurricane season was almost upon us, when no skipper wants to be caught at an exposed dock. I remembered that the last big hurricane had brought a 15-foot sea across the flats of the lagoon and on through the docks where we were presently tied up.

I examined the bronze shaft and found it bent, as I had feared. Resolutely gathering up my tools and scuba gear, I disconnected the coupling from the end of the shaft, slid off the coupling, and prepared a wooden plug to prevent water from pouring in the 1-inch opening. I fervently hoped the threatening thunderstorms would hold off until our new shaft arrived. Joan flew the bent shaft to Miami and went to a machine shop to order a new one.

Six days later, when Chalk's seaplane brought the replacement, one of my divers, Roger Haydock, helped me to replace the shaft underwater, whereupon we learned that a new stern bearing was also needed. I breathed a sigh of relief when one was found in the spares. Keeping a 9-ton sloop afloat with a wooden plug for a week during hurricane season was tough on the nerves.

As if these practical tests of our determination weren't enough, other challenges awaited us at Bimini. Archaeologists in the field—especially those far from urban centers—have a unique set of problems. Weather, snakes, treasure-hunters, and insects can all make life miserable for the dirt archaeologist. But in fair weather they can at least usually count on full days on the site. Such was not the case for us. In addition to normal distractions, we had to contend with fishing boats passing overhead at high speed and running down our buoys, tourists snorkeling nearby, and even a film crew shooting a pornographic movie!

Foremost among our difficulties was the constant problem of visibility. The best time for underwater work occurs during the flood tide; on the ebb, tiny biological debris from the banks can shut down the visibility in minutes. One instant you can see 60–100 feet away; in the next, only 10–15 feet in what resembles an underwater snowstorm, as the sun illuminates millions of tiny particles. Thus, although the best visibility was between 9 A.M. and 4 P.M., a rain shower or ebb tide might cut our time to little over an hour of good working conditions. Another factor adding to our difficulties was the immensity of the site. We learned to mark carefully the specific locations of even minor check points; otherwise, they could easily be lost in the vast waters and shifting sands.

Certain species of poisonous reef fish were avoided by learning to recognize them, followed by caution. Now and then, through carelessness, someone suffered the pain of an embedded sea-urchin spine, and fire coral occasionally gave an unwary diver a burning, itching sensation where his body had brushed against it.

Those who imagine that working underwater in the tropics is glamorous will develop a more sober perspective after meeting a barracuda. His mouth open with wicked teeth ready, breathing slowly, all 3–5 feet of silvery body represent maximum engineering efficiency for speed, cutting, and slashing. Before getting in the water, we had to dull down all bright, shiny metal that might be a part of our gear, since barracuda will mistake such glints and glitters for edible minnows. Even so, a number of times I was surprised by single barracuda or groups suddenly appearing on my flank. Despite the number of times I have encountered these fish—knowing full well that these creatures are mostly curious and that there are only thirty documented attacks—my gut reaction to their sudden appearance has always been the same. Regardless of your degree of experience, caution is advised. The best retreat is made by moving slowly and deliberately away—always facing them.

Even when you begin to feel at home in an underwater site, you can be surprised. While working on the tape-and-compass survey one day, Gary Varney was holding the end of the tape on the edge of one of the megalithic blocks. Suddenly a moray eel seized his finger and tried to pull his arm under the block. Even though it was a small eel—we had seen two 6-foot green moray eels on the site—Gary had to struggle to free himself and came away with a badly lacerated finger. We cleaned and dressed his wound immediately, but within the hour the lymph gland under his arm nearest the wound began to swell. Fortunately we had antibiotics aboard. Moray eels' mouths are incredibly filthy, and wounds made by them must be attended to at once to avoid serious infections.

In surprising contrast, sharks were much less threatening during our project. By diving to avoid feeding times, (again between 9 A.M. and 4 P.M.) under good visibility, we increased our chances of safety. In my Bimini experience, surprisingly, it has been the smaller sharks that were the most aggressive. The most hostile one I met was only about 2–3 feet long, moving quickly below me at a depth of about 18 feet, with erratic side-to-side motions of his body. Suddenly he tore savagely into a school of small reef fish, then flashed on. The capability of violence this small specimen displayed gave me a healthy respect for the potential of larger sharks. Because sharks have some unknown mechanism for perceiving fear, I caution anyone entering shark-occupied waters to be sure they can control their actions. If confronted by sharks or barracuda, it is vital to avoid the violent splashing motion of a wounded fish.

But the most preposterous hazard came from the activities of some mainland university students who were running some experiments in marine biology concerned with sharks. One day we found them over the Paradise

Point site as we arrived to begin the day's diving. They had electronic devices lowered into the water from their rubber boat. When we asked them what they were doing, we were told that they were trying to attract sharks! Appalled, we asked them to move their laboratory elsewhere because we had work to do which couldn't be moved. To our relief, they agreed, and we began our day's work. The very next day, however, we found them once again on the site, this time with pieces of bloody meat in the water! The ensuing conversation finally convinced them that while their activities might be important to science, they threatened our safety. From time to time we saw them afterward, but not at the site. Operating elsewhere, they were doubtless improving our underwater habitat.

Then at one point in the summer, after sundown, three of us were taking *Makai II* to Miami from Bimini. Twelve miles off Bimini, we had been sailing through a series of rain showers, some of them squally. Breaking out of one, we saw a vessel 500–600 yards dead ahead. From its running lights, it seemed at least 100 feet in length, but it is easy to be wrong at night when judging such matters. The vessel flashed a blinding spotlight at us. I immediately tried the VHF calling channel 16, which is also used for distress calls and is monitored by most vessels making the crossing to the United States. Although there was no response, I knew that our radio was working because its output indicator lamp was lit, and the boat had an excellent high-gain antenna.

Rather quickly we began to close with the unidentified vessel. Finally, to avoid the danger of a collision, we put the boat over on the other tack. Soon the other vessel maneuvered to reestablish the collision course. I then tried the other radio-telephone on a lower band, using 2,182 kilohertz, the international distress frequency. By this time, all sorts of wild thoughts were running through our minds. Was this a Bahamian gunboat involved in the lobster war between Florida lobstermen and the Bahamian government? Was it a vessel bent on piracy? A few months before the expedition began, I had read a story from Coast Guard sources about the unexplained disappearance of several hundred yachts within the previous two years. All of the missing vessels had long-range capability. Generally the crews were never seen again, although a few turned up in small boats with tales of being boarded in the middle of the night by armed men, followed by gunplay. The official position was that these vessels were probably lost to piracy undertaken to support drug smuggling.

I was wondering if we should break out our small arms and try to defend ourselves, when we were again forced to tack to avoid an imminent collision with the strange vessel. Moments later a small boat came up quickly from astern, then crashed against our hull on the starboard quarter. Then and only then did I recognize that we were dealing with the U.S. Coast Guard! As the boarding officer stepped on deck, he told us that they were conducting a safety inspection. I couldn't believe it! Here we were, under sail in international waters, hazarded by close maneuvering, tacking in and out of rain squalls, all compounded by their failure to maintain a radio watch on the two mandatory frequencies! I'm afraid I was rather sarcastic with the boarding officer. He went through the usual bit of checking the ship's papers, asked to see the legally required safety equipment and so on. Ironically, as he left, he forgot his own lifejacket. I threw it after their departing boat. Since that time I have wondered how often the Coast Guard gets fired upon if they handle other boardings as they did ours.

6

MORE DISCOVERIES—AND SOLID EVIDENCE

All summer, one question was uppermost in our minds: What was the original function of the Bimini Road? Would we find anything relating it to a sacred site?

True, as our work continued, the evidences of artificial patterns in the huge blocks continued to accumulate, and the same evidence also seemed to favor the presence of a megalithic site. But then one day, while diving on the Paradise Point site, I observed a strange grouping of three stones. I had passed over them dozens of times without noticing the pattern, and now wondered how I could have missed it. The three stones formed an arrow configuration, and when related to the nearby base of the chockstone, the group seemed to point toward the east.

Encouraged by this hint of order in the ruins, I swam

northeast up the main lead where, to my surprise, I observed two long, thin, nearly cylindrical stones at approximately the same distance as between the arrow and the chockstone. They looked very much like obelisks, or standing stones, although now, like all of the stones on the site, they were resting in a horizontal position. Because they tapered to one end and rested on several smaller stones, it is conceivable that they once stood in a vertical position. It seemed unlikely that this geometric layout oriented to the equinox was an accident of nature.

Now, more clues began to fall into place. I recalled that the number five had been found to be emphasized in two places: in one portion of the site we found an extensive line of fourteen single stones about 4½ feet square, which gave way to a pavementlike section five stones wide, each of them about 8 by 11½ feet. This very impressive area ran over 105 feet to the northeast.

In another section, which John Parks felt was highly evidential of human intervention, five large stones were found side by side with five small ones, changing to a pattern of six pairs of stones equal in size. This stressing of the numbers five and six in the layout brought to mind John Michell's connection of these numbers with known megalithic sites.

Late in the summer, John Steele led a party that discovered another possible "megalithic link"—they found fragments of huge stones indicating that the shorter arm of the J could have once extended a considerable distance. This suggests that the original design could have been shaped like a horseshoe or a hairpin, a feature also found at Stonehenge. (Like Stonehenge, the site at Bimini also has a northeasterly orientation.)

The arrival of an unexpected visitor soon revealed that only a few hundred yards west of the Road lay another possible clue to early human presence. Jacques

Meyol, a diving expert and a friend of Dr. Valentine's, wanted us to see a puzzling trench close to the Road site, which he and Dr. Valentine had come across in 1970. When John Parks joined us for a close inspection of the trench, he concluded that it was not a natural fault in the seabed. No fractures were in evidence at the corners of the flat-bottomed trench, which seemed to have been cut out of the solid limestone of the seafloor.

Its width was just over 4 feet at the top and almost 20 inches at the bottom. The trench wall closest to the beach was nearly vertical, while the seaward one flared out at about 45 degrees from the vertical. In cross section, it was therefore asymmetrical, apparently a deliberate construction. Overall it ran about 108 yards and 25 inches deep. From its northern end, the magnetic bearing to the Rockwell estate on Bimini was about 165 degrees and to the smallest of the Crossing Rocks (the seaward one), about 205 degrees. The northern 260 feet of its length pointed in the direction of 30 degrees azimuth (magnetic); the southern 66 feet in the direction of 10 degrees. The shorter southern section overran the junction like a pocket cut made with a power saw. This abrupt change of direction, in John's opinion, was also difficult to reconcile with a natural fault. The more we looked, the less plausible were any of the ordinary explanations. All in all, we found the trench another intriguing mystery of the Bahamian waters.

For many weeks we worked on the geological problem and the site survey, finally coming to the last week of the expedition. In the tropical heat of late summer in the islands, everyone was tired. On August 7, we were all checking out a new pattern in the blocks when suddenly, 20 feet below, I spotted what appeared to be a human-made shape. Diving down, I found on closer inspection that the stone object was definitely human-made. Only about 6 inches of one edge was exposed, and

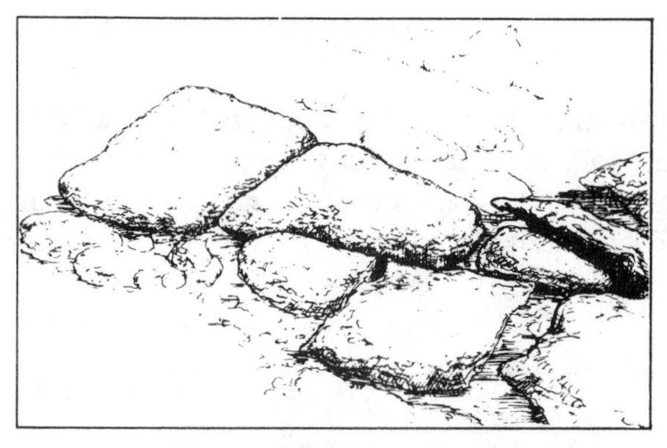

"Arrow" stones. (Courtesy John Parks.)

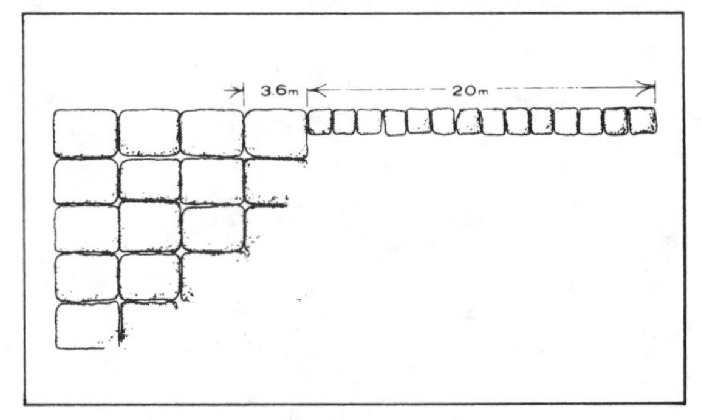

The fourteen-stone–five-stone pattern. (Courtesy John Parks.)

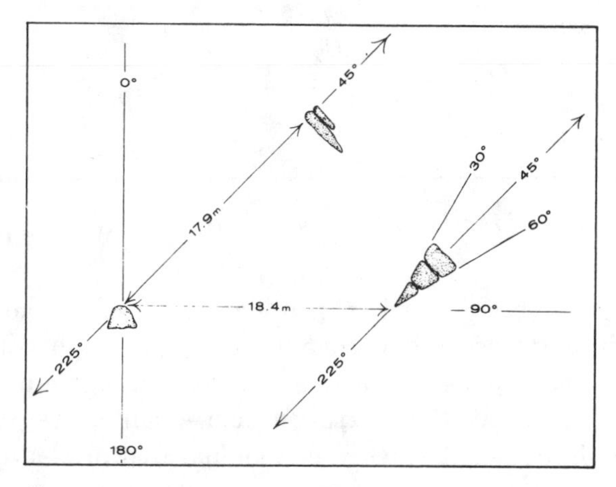

Geometrical layouts at Paradise Point site. (Courtesy John Parks.)

the minimal marine life on this part suggested that the current had only uncovered it very recently. Within minutes I got the others over to have a look, then photographed it before raising what proved to be an ancient building block.

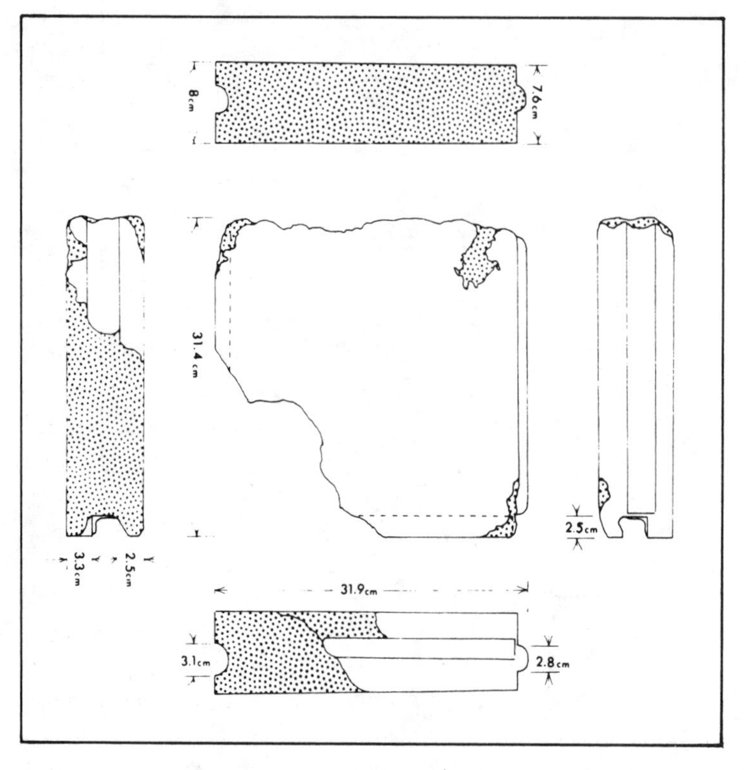

Dimensions of the building block. (Courtesy John Parks.)

Despite the gathering thunderstorm overhead, we were jubilant as we saw that it had a very sophisticated tongue-and-groove pattern and was a fragment of a larger piece. With careful measurements, we determined that its flat sides were not parallel, and that its varying thickness of about three inches suggested a sloping wall or molded brick, and we hoped that it might have been fired and thus datable by available scientific methods. The chert and limestone elements evident in its composition indicated a sandstone and limestone mixture not found in the Bahamas. Nothing definite could be said about it until we could date it and, if possible, assign it to a known culture pattern. But as we studied the weathering of the block and recognized the sophisticated jointing pattern, it was obvious that it was extremely old, and did not relate to the simple culture of the local Lucayans. Could it be a relic from the site of another ancient culture?

Only six days later, Gary Varney, who was a dowser as well as a diver, made a discovery at a considerable distance from our regular area of exploration. Gary's dowsing had picked up what would be one of our most important finds. By now I was letting my emotions have a freer rein, and excitedly dove down where Gary led me. Here we found a stylized head which we estimated at 200–300 pounds. It seemed more like an animal head than a humanoid one, particularly from its left side. Rolling it over, I knocked a small chip from a bottom corner. When we surfaced, the chip proved to be a beautiful white marble! No marble is native to the Bahamas, and this chip had been eroded at least an eighth of an inch by the destructive processes of the sea. Recognizing the significance of this find, I realized that to raise it, we would have to get special permission from the government. Had I known the trouble it would later entail to obtain, I would have gambled on the political hassle of raising it there and then.

Whatever the culture pattern eventually assigned to the Paradise Point site, I now felt that we were on our way to completing the first phase of our search. The discovery of two human-made artifacts with deep weather-

The author inspecting the ancient block on board after its retrieval. (Courtesy Joan Zink.)

The marble head in its original position 20 feet below the surface.

ing from bioerosion was a real break for the expedition. Now we would try to date the building block by absolute methods; we would also check to see if it had any relation to known pre-Columbian cultures. Work on the head would have to wait, but we would circulate its photographs among pre-Columbian experts for opinions. Until the head was lifted, the marble chip from its base would be held for safekeeping.

As we finished for the season, we were confident that we were dealing with an archaeological site of some ancient, unknown megalithic culture, and producing hard evidence to support the speculations of others who had

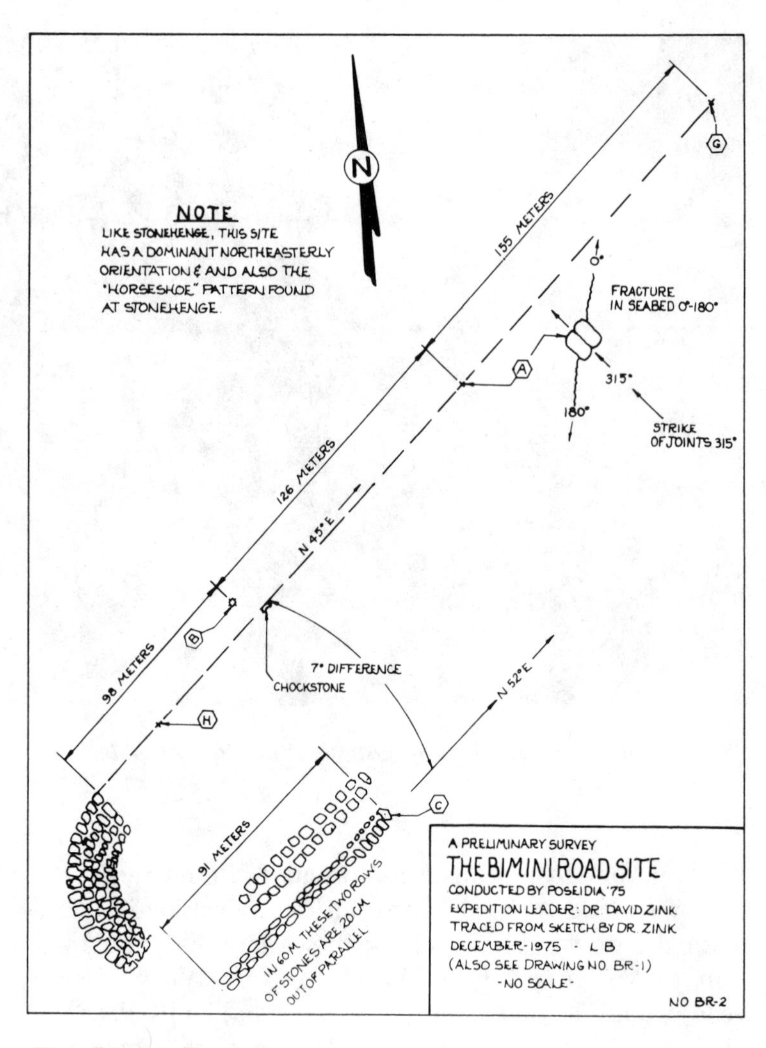

The Bimini Road site.

previously written of the finds off Bimini. We had already concluded that there was no rational basis for calling the Road site a road. As John Steele pointed out, the hairpin turn or curve—if indeed a part of a road—belonged in mountain terrain, not on the level seabed where we found it. Although the reversed J shape did suggest a jetty or harbor works, its present form lacks the height to be effective in this function. (Possibly, of course, it once included more courses of stones and served various functions including harbor protection, though later clues continued to suggest a sacred function for the site.)

John Steele, familiar with European megalithic sites, commented on the size and weight of the blocks, the hints of certain sacred numbers in the blocks' patterns, and the possible orientation to the solar equinox position in the layout. Added to our finding of the freshwater spring near the site visible in an aerial photo in 1974 (another indication of a sacred site), this all tended to bring Bimini into the arena with New Hampshire's Mystery Hill as a contender for one of the earliest megalithic structures in the New World. In his archaeologist's report after the expedition, John confirmed: "I have come to the conclusion that it is definitely an archaeological site of megalithic construction. There is a plausible inference that it also had a sacred function."

PART II

THE QUEST CONTINUES

7

ATLANTEAN THEORISTS

Back in Virginia, I experienced a great deal of emotional and mental turmoil. Aside from the project's complicated aspects and controversial implications, the summer of Poseidia '75 had brought me to a crossroad in my life: soon I would have to decide whether to return to university teaching. Any meaningful research into new aspects of the Bimini enigma would obviously be a full-time task—new problems were already developing at an alarming rate. I had to consider the pros and cons of continuing this investigation.

True, I was beginning to feel I might be able to make a real contribution to the problems the Bimini site posed. Yet in the tree of academe, it was professionally dangerous to go out on such a rickety limb. Those Bahamian islands could absorb a lot of time and money before giving up their secrets. I had learned enough to suspect strongly that we were dealing with an ancient site, but I knew it would be hard to prove the fact to a skeptical world, and still more difficult to determine who had inhabited it and when. And, I asked myself at low moments, could I actually solve any of the countless puzzles of prehistory? Would the possible rewards of this quest justify my leaving university teaching and possible recognition as a Victorian scholar? On the other hand, academe itself might be left behind in the long run, as students slowly began to recognize all the unexplained anomalies in prehistory. Perhaps it was better to seek further answers and insights by personal investigation out of the library and in the field.

Eventually I began to realize that my decision could never be made on a purely rational basis. In many solitary hours of inner debate, I forced myself to reconstruct the most important aspects of the evidence and the thinking that had sent me to Bimini in the first place. What finally emerged was a strong conviction that it was both important and necessary to proceed with the investigation. Now, I realized, it was time to head back to the libraries to dig into the known and suspected findings on both Atlantis and Bimini, from myth through science. This phase of the work, I knew, would help me get through the long winter months and prepare me for a new Poseidia search in 1976.

Much of Plato's legend (see appendix A) is obviously beyond verification today. Yet as I thought about it, I realized that some aspects were more susceptible to investigation than others. For example, among the trees and abundant fruits which Plato described on the land were those yielding "drinks, meats, and ointments." Other sources have reinforced my conclusion that he was referring here to the coconut, which is not native to the Mediterranean, and was unknown to the Greeks until about a century after Plato's time. I decided to begin a series of notebooks in which I could record facts which seemed to reinforce or dovetail with each other.

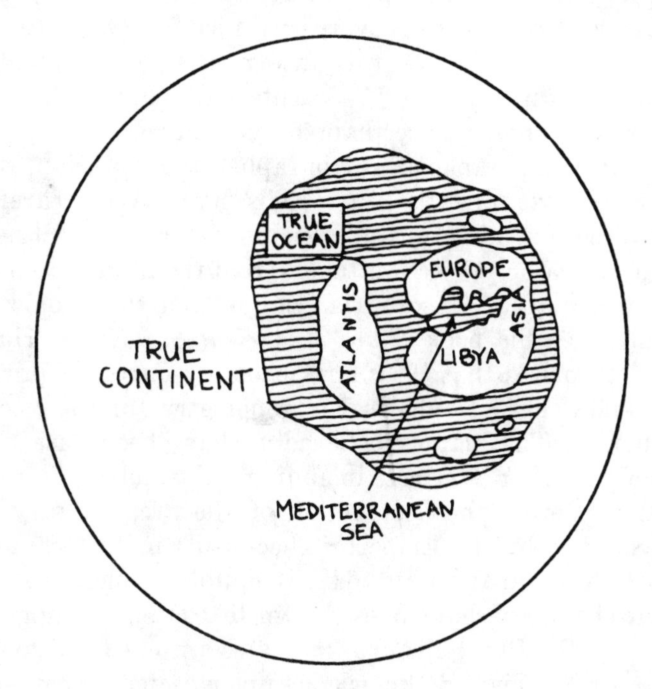

Plato's conception of the terrestrial sphere.

Continuing on with Plato's story, I found reference to a particularly important factor: the *location* of Atlantis. In the *Timaeus,* Critias says that voyagers going west from Atlantis "had access to the islands and from the latter to the opposite continent, which is located at the edge of the real ocean." Many have taken this to be a reference to the Pacific. How could Plato, supposedly ignorant of the existence of the Americas and the Pacific, write with this geographical awareness?

The answer may lie with the ancient Piri Re'is chart,

compiled by a Turkish admiral around 1513—twenty years after Columbus' first attempt to reach America. Amazingly, it shows South America and may also reveal outlines of land under the Antarctic ice. Since the coast of Antarctica has been covered with ice for 20,000 years, one can well ask how this information, only recently confirmed during the 1957–58 International Geophysical Year, was known more than 450 years ago.

Charles H. Hapgood, cartographer and historian, was asked to examine the Piri Re'is map. After careful study he and his students concluded that it was based on a Mercator projection taken from the intersection of the meridian of Alexandria, Egypt, and the Tropic of Cancer. In his book, *Maps of Ancient Sea Kings,* Hapgood theorizes that the charts used as sources required the knowledge of spherical trigonometry for their construction. The map is believed to have been based on twenty earlier sources including Alexandrian Greek charts. Before the destruction of the great library at Alexandria by the Emperor Theodosius in A.D. 389 and the Caliph Omar in A.D. 642, it doubtless included ancient charts, perhaps ones known to Plato. If Hapgood is right, the Piri Re'is map indicates a prehistoric civilization with scientific knowledge unsuspected by contemporary man.

Although I had read a great deal relating to the Atlantis legend during my teaching days, I knew I would have to delve even deeper to discover connecting links between Atlantis and the Bimini Road. My small black notebooks continued to fill up as I spent longer and longer hours at my research. Plato's legend had indicated a great deal of controversy; now I wanted to learn what conclusions—if any—his successors had reached.

Even granted that the legend of Atlantis has always intrigued the human imagination, I was amazed to find that the topic has been treated by some five thousand works in twenty languages! In the last few years alone, interest has grown to an incredible degree. Modern Atlantean theorists are the latest in a long line of such thinkers who (in Aristotle's case, for example) go back to Plato's own era. In the past century, over a dozen major theories of Atlantis have surfaced. Many of these are nationalistic in thrust, locating Atlantis in the writer's own part of the world. But as I reviewed their work, I found that all these theorists fell into one of three distinct groups:

1. Those who believe that Plato contrived the legend himself. The earliest skeptic of Plato's account was his own former student, Aristotle, who himself wrote of an island in the Atlantic known to the Carthaginians as Antillia. In reference to Plato's Atlantis, he quipped, "He who invented it also destroyed it." The latest and best known skeptic is L. Sprague de Camp whose book, *Lost Continents,* maintains that Atlantis is merely a literary theme Plato devised to continue the moralizing of his *Republic.*
2. Those who would edit Plato, either by moving the site of Atlantis out of the Atlantic or by dividing his measurements by a factor of ten, or what have you. This group is represented by writers like Velikovsky and by the Greek seismologist Galanopoulous, who claimed in 1960 that he had found Atlantis on Thera in the Aegean. (see appendix C.)
3. And, finally, those who thought that Plato was giving an honest account of events that may really have happened. In the course of my researches, I began to find myself supporting this latter group, which has as its pioneer spokesman Ignatius Donnelly (1831–1901), who was one of America's most

> learned congressmen. After years of work in the Library of Congress, he published his book, *Atlantis: The Antediluvian World,* in 1882. I feel this work initiated our modern study of the Atlantis question and outlined the major problems for future research and investigation.

Donnelly built on the work of earlier writers who considered the Azores and the Canary Islands to be remnants of Atlantis and others who had argued that the Mayas were descendants of the Atlanteans. During the ongoing rediscovery of the Mayan culture in Mexico's Yucatan, in British Honduras (now Belize), Honduras, San Salvador, and Guatemala, many have commented on the strong architectural parallels between the Mayan and the Egyptian culture: step pyramids, columns, obelisks, stelae (upright slabs of stone with inscriptions or sculpturing), the use of hieroglyphics as ornamentation, bas-reliefs, and a strange absence of the true arch. Both cultures had the corbelled arch, as did ancient Mycenaean Greece.

In addition to the architectural parallels, Donnelly found significant similarities between Egypt and Central and South America: early use of the solar calendar, sun worship, mummification, and pyramids. Donnelly believed that Atlantis represented the first civilization on this planet, that it was the center from which these characteristics were diffused, that its far-reaching colonies extended from the Atlantic base to as far as Central Asia, and that its destruction was historically factual. He set forth his position in thirteen propositions:

1. There once existed in the Atlantic Ocean, opposite the mouth of the Mediterranean Sea, a large island which was the remnant of an Atlantic continent, and known to the ancient world as Atlantis.

2. The description of this island given by Plato is not, as has been long supposed, fable, but verifiable history.
3. Atlantis was the region where man first rose from a state of barbarism to civilization.
4. It became, in the course of ages, a populous and mighty nation from whose overflowings the shores of the Gulf of Mexico, the Mississippi River, the Amazon, the Pacific coast of South America, the Mediterranean, the west coast of Europe and Africa, the Baltic, the Black Sea, and the Caspian were populated by civilized nations.
5. It was the true Antediluvian world, the Garden of Eden, the Garden of Hesperides where the Atlantides lived on the River Ocean in the west; the Elysian fields, situated by Homer to the west of the Earth; the Gardens of Alcinous (grandson of Poseidon and son of Nausithous, King of the Phaeacians of the Island of Scheria); the Meomphalos, or Navel of the Earth—a name given to the Temple at Delphi, which was situated in the crater of an extinct volcano; the Mount Olympus of the Greeks; the Asgard of the Eddas, the focus of the traditions of the ancient nations; representing a universal memory of a great land where early mankind dwelt for ages in peace and happiness.
6. The gods and goddesses of the ancient Greeks, the Phoenicians, the Hindus, and the Scandinavians were simply the kings, queens, and heroes of Atlantis; and the acts attributed to them in mythology, a confused recollection of real historical events.
7. The mythologies of Egypt and Peru represented the original religion of Atlantis, which was sun worship.

8. The oldest colony formed by the Atlanteans was probably in Egypt, whose civilization was a reproduction of that of the Atlantic island.
9. The implements of the Bronze Age of Europe were derived from Atlantis and the Atlanteans were also the first manufacturers of iron.
10. The Phoenician alphabet, parent of all the European alphabets, was derived from an Atlantis alphabet, which was also conveyed from Atlantis to the Mayas of Central America.
11. Atlantis was the original seat of the Aryan or Indo-European family of nations, as well as of the Semitic people, and possibly also of the Turanian races.
12. Atlantis perished in a terrible convulsion of nature in which the whole island was submerged by the ocean, with nearly all of its inhabitants.
13. A few persons escaped in ships and on rafts, and carried to the nations east and west the tidings of the appalling catastrophe, which has survived to our own time in the Flood and Deluge legends of the different nations of the Old and New Worlds.

The nearly universal incidence of flood stories around the world and widely distributed legends of destructions by fire and water provided the background for Donnelly's theorizing, as they later would for Velikovsky. The Mayas and the Aztecs, for example, tell of four destructions; other cultures speak of as many as seven. Donnelly was probably the first to focus upon the peculiarities of the Basque people of the Spanish Pyrenees, whose language cannot be related with certainty to any other language or group of languages. (Later, I would find other references to this area as an Atlantean colony.)

Many anthropologists would now condemn Donnelly

as an extreme diffusionist—all theory and no facts. They would explain the similarity of cultural traits as part of the psychic unity of mankind, a concept originally formulated by Adolf Bastian. According to this theory, similar environmental and cultural conditions tend to produce similar human responses and similar solutions to common problems. Thus, parallel developments in widely separated cultures might be expected to occur independently and be "discovered" only when communications were finally established between the cultures in question.

Let us assume, for example, that two Stone Age cultures, innocent of the engineering principle of the true arch, needed openings in their stone walls. They might devise the post and lintel doorway—but they could choose another approach. By piling the stones up in a wall, gradually moving the stones of each new row toward the center of the opening—all without outside influence—thus would achieve what is now called the corbelled arch. Those who favor independent invention would explain by this process the similarity of other transatlantic cultural traits.

But for the record, it should be noted that diffusionism is alive and well in a recent book by linguist and historian Cyrus H. Gordon, *Before Columbus: Links between the Old World and Ancient America.* Professor Gordon hypothesizes a Bronze Age seafaring culture that, in the pre-Columbian era, diffused Old World culture across the Atlantic to the New World. Gordon's work reinforces Donnelly's diffusion theory by suggesting that prehistoric peoples were more far-ranging in their travels than previously realized. (Twentieth-century anthropologist and explorer Thor Heyerdahl was sufficiently impressed with the notion of Egyptian crossings of the Atlantic to undertake such a voyage himself in his reed boat *Ra.*)

The impact on Atlantean studies of Donnelly's pioneering investigation was obvious to me in the writings of Lewis Spence (1874–1955). Until his death, Spence was the foremost student of Atlantis in this century. A Scottish mythologist, Spence wrote nearly fifty books, four of which are directly related to Atlantis: *The Problem of Atlantis, Atlantis in America, The History of Atlantis,* and *The Occult Sciences in Atlantis.*

The first of these books, reprinted in 1972 as *Atlantis Discovered,* is regarded by the Atlantis skeptic L. Sprague de Camp, writing in about 1970, as "still about the best pro-Atlantis book published to date." One reason for his opinion may be the fact that Spence claims no more than a Stone Age culture for Atlantis, about on par with the Mayan and Incan cultures. (These later cultures, of course, had rich inner lives; the Mayas, at least, had an evolved astronomy and mathematics. The limits of the term "Stone Age" have been recently recognized and it is therefore much less frequently used in professional circles.)

Spence, with the advantage of a more recent anthropology and paleontology, located Atlantis in the Atlantic and maintained that it sank by degrees. First it was a continental mass that began to break up late in the Miocene geological period (beginning about 25 million years ago), finally separating into two smaller masses, Atlantis and Antillia (located by Spence in the vicinity of the present West Indies). These island-continents continued to exist until about 25,000 years ago, when Atlantis further disintegrated, and was at last totally destroyed about 10,000 B.C. In Spence's theory, some parts of Antillia still survive in the West Indies, although he does not draw specific attention to the Bahamas. This piecemeal destruction, while it runs counter to Plato's "overnight" version, is more compatible with geophysical evidence and also with the accounts from Edgar Cayce.

Spence equated two of these destructions with three cultural waves that seem to have reached Europe from the west. These are the Cro-Magnon or Aurignacian culture (dated in Spence's day at about 25,000 years ago); the Magdalenian or Cro-Magnon culture of about 16,000 years ago; and the Azilian-Tardenoisian of about 10,000 years ago. The respective current dates for these cultures are 34,000–27,000 B.C., 15,000–10,000 B.C., and 8000–6000 B.C. In Spence's opinion, these cultures left Atlantis one after another because of natural catastrophes and made their way to Europe. In particular, the Cro-Magnons of the Aurignacian and Magdalenian cultures seem to have been highly developed before reaching the Iberian peninsula, since evidence of earlier Iberian development is lacking. An eastern origin seems to be precluded because Neanderthals were located there. Cro-Magnons appear not to have been descendants, but instead apparently conquerors, of the Neanderthals.

The Aurignacian and Magdalenian cultures produced the stunning cave paintings found in France and Spain. Their extraordinary sense of anatomy and perspective in drawing—the ancient artists used even the curve of the cave wall to enhance their art—suggest a surprising modernity for Stone Age cultures. These cave paintings also reveal a central concern of Cro-Magnon peoples—the bull. Bulls, it will be recalled, were sacred animals in Plato's version of Atlantis. They are also found elsewhere: in the Mediterranean there was the sacred bull of Egypt, Apis, and the bull of the Minoans on Crete.

The late Egerton Sykes, a fellow of the Royal Geographical Society and Ignatius Donnelly's editor, succeeded Spence as a foremost researcher of Atlantis. Sykes brought great depth of learning in the classics to the Atlantean problem. He was the editor of *Atlantis,* an interesting publication now in its twenty-eighth year, whose recent issues have focused on the Bermuda Triangle and Bimini. The journal is an excellent source of knowledge about the work of Atlantologists around the world.

Also among the pro-Atlantis books is Charles Berlitz's *The Mystery of Atlantis.* I found it a helpful survey of the various theories that remove Atlantis from the Atlantic and locate it in North Africa, Scandinavia, or elsewhere. Charles Berlitz, grandson of M. D. Berlitz, who established the worldwide Berlitz language schools, is himself in command of thirty languages and is a scuba diver. He is a good example of today's researcher who realizes that library investigation must be complemented by personal field study of various sites. The strength of Berlitz's book is, as might be expected, in the area of linguistics.

Two examples from Berlitz's work illustrate the role of language in the Atlantean problem. First, the Aztecs referred to their original homeland as Aztlan—an echo of the original sunken land? Also, in Aztec, *atl* means water. Across the Atlantic, in the Berber tribe of North Africa, the same word has the same meaning. These and other parallels suggest to Berlitz and others the possibility of a common origin in Atlantis.

In the following example from Berlitz's book, the idea of chance origins in widely separated linguistic islands of sounds of amazing similarity for the word *father* seems quite *im*probable.

Basque (suspected Atlantean fossil language)—*aita*
Quechua—*taita*
Náhuatl (Aztec)—*tata* or *tahtli*
Zuni—*tachchu*
Welsh—*tàd*
Fijian—*tata*
Turkish—*ata*
Maltese—*tata*

Roumanian—*tata*
Samoan—*tata*
Dakota (Sioux)—*atey*
Seminole—*intáti*
Tagalog—*tatay*
Sinhalese—*thàthà*

For Berlitz, these parallels imply an earlier diffusion of language than is currently recognized. Of course, the antidiffusionists argue that such developments in language are a matter of chance. In the area of linguistics, as in other disciplines brought to bear upon a problem as difficult to pin down as Atlantis, we are not talking about proof, of course. Instead, we can talk only about the more or less probable.

In researching contemporary Atlantean theory, one of my surprises was the discovery of a strong interest among a number of Soviet academics. Not only is the subject taken seriously there but Atlantean research is being pursued with an impressively broad outlook. Doctor of Chemical Sciences N. F. Zhirov, writing almost twenty years ago in Egerton Sykes' journal *Atlantis,* said that the study required "scholars with adequate encyclopaedic perspective" since "proof of its reality consists of a complexity of innumerable small facts and observations connected with diverse branches of science." I was especially intrigued by a 1961 book published by the (Russian) State Publisher of Children's Literature entitled *In Search of the Lost World: Atlantis;* its twenty chapters were as scientifically sophisticated and comprehensive in its treatment of the topic as many adult books in the West.

The background I received from the above theorists' work gave me an invaluable opportunity to confirm or reject elements of my own slowly developing hypothesis. However, my attitude was also being shaped by evidence from other areas. One source of information that had a tremendous impact on my theory and brought an entirely new approach to the problem was the American psychic, Edgar Cayce.

As I read more and more of his material, I found myself highly impressed with the internal consistency of his readings on Atlantis. Within the mass of information he gave are medical and psychological readings that have demonstrable, practical utility for gains in health and personality growth, factors which make Cayce's account of Atlantis difficult to ignore. Even his originally improbable claim of over 10 million years of human presence on the planet begins to seem more plausible as science pushes back the past. (The work of the Leakeys in Africa has recently put human origins at at least 3.75 million years ago.)

The Cayce readings trace Atlantis from its beginnings through its golden age and then on through three increasingly drastic earth changes, including the one that finally destroyed the civilization. These occurred at about 48,000 B.C., 28,000 B.C., and 10,700 B.C.

Cayce's Atlantis featured great stone cities; modern communications including electronics; land, air, and undersea transportation; the neutralization of gravity; and the harnessing of solar energy by means of the great crystals, or "fire stones." This exotic technology is probably the more bizarre aspect of Cayce's account. What was accomplished with this high technology, however, begins to look like a script for the decline of our own present civilization. From its golden age, Atlantis deteriorated, spiritually and morally, until the great forces of nature available through the crystals began to be misused. This technological abuse eventually caused at least one of the cataclysmic events, that which caused the final destruction through violent earth changes.

The deterioration of Atlantis was a total decline of a

great civilization; in its last phases, at least in moral and psychological terms, it suggests the latter days of the Roman Empire. Atlantis' original spiritual awareness and her citizens' selfless discipline gave way to increasing materialism, sexual perversion, exploitation, and slavery. The decline was marked by a sharpening conflict between the Sons of the Law of One and the Sons of Belial. (The reader may recall the allusions to the Sons of Light and the Sons of Darkness found in the Dead Sea Scrolls.) The Sons of Belial ultimately came to practice human sacrifice, sexual license (including psychic control of others for sexual purposes), and finally, the misuse of the forces of nature. The fire stones, which earlier had been used for healing, began to be employed in punishment and torture. As I read these details, I wondered if such abuse and the psychic exploitation of others might represent the beginnings of black magic.

Setting aside the paranormal elements and the sophisticated technology, the moral decline parallels Plato and other accounts of more recent vintage from psychic sources, such as that of the Theosophical Society started in the last century by Madame Blavatsky. The Atlanteans, many of whom Cayce saw incarnated in the present century, were extremists, either desiring greatly to serve humanity, or to exploit others. Cayce saw that many had individual or group karma involving exploitation of others, a debt which has yet to be worked out through love and service.

Intrigued by Cayce's readings, but feeling I now needed grounding in a more concrete area, I turned to geological evidence for the existence of Atlantis—and learned, to my surprise, that this science's foundations were far less solid than I had expected.

8

CLUES FROM AN UNSTABLE EARTH

Geology is now dramatically readjusting long-held conceptions of the earth's history. Over the past thirty years, it seems the solar system's age was pushed back to 4.5 billion years. Then, radioactive material within the earth's core was found to be a cause of movements within the earth's crust. Electrical currents in the earth's core were identified as the most likely source of the earth's magnetic fields. The theory of continental drift, now updated as "plate tectonics," in which slabs of lighter material (the continents) break up and drift away from each other, is now recognized as a reality. The ocean floor is younger than earlier suspected, due to the drifting apart of continents and the simultaneous creation of new seafloor as the continents separate. Finally, the previous idea of only four ice ages has given way to a theory that one occurs every 100,000 years. All this has added up to a much more unstable planet where awesome forces, continually at work, have sometimes produced earth changes on a scale not previously recognized.

An article I had first read three years earlier by Dr. Cesare Emiliani, marine geologist at the University of Miami, refreshed my awareness of the ocean-bed core samples taken by the *Glomar Challenger,* an oceanographic research vessel. It was discovered that a 1,500-mile-long ridge under the Indian Ocean, now over 1 mile deep, had once been above water—"an island chain with swamps and lagoons." Other deep cores, with recent deep sea sediments in their upper layers, included sediments usually found in shallow water *and* dry land. Emiliani interpreted the discoveries to mean that, just as underwater debris can be raised "to great heights above sea level to form mountain chains . . . the reverse apparently also happens on a scale not yet foreseen."

During the same *Glomar Challenger* voyage, it was also discovered that 6 million years ago, the Mediterranean Sea was a desert 10,000 feet below sea level. Five and a half million years ago, the sea broke through the land bridge connecting the Straits of Gibraltar, and marine life from deep in the Atlantic poured in with the torrent which, it was estimated, took about 1,000 years to fill the basin. (I suddenly recalled that Herodotus had held that the Straits of Gibraltar had once been joined by land.)

The geology of Atlantis, as described by Plato, includes hot and cold springs (which suggest a volcanic terrain), and stones whose colors were white, black, and red. Rocks of these colors have been found in Atlantic islands, such as the Azores. Structurally, the Azores appear to be remnants of the mid-Atlantic Ridge, now deep in the Atlantic. This ocean bottom feature has been seen by some as the remains of the sunken Atlantis. The Azores continue to exhibit extensive volcanic activity, as do the Canaries and other islands associated with this ridge.

Evidence of the ridge's instability became evident in an incident in 1898. Five hundred miles north of the Azores, a cable-laying expedition was working in 1,700 fathoms of water. During an attempt to raise the cable, grappling hooks brought up a sample from the ocean floor. The sample was tachylite, a vitreous lava which is formed above water, not beneath it. This type of lava is also supposed to disintegrate in sea water in about 15,000 years. If these assumptions are valid, we can infer that this particular tachylite sample formed from a volcano that was above water about 15,000 years ago and subsequently sank to its present depth. These findings add substance to the claim that within the past 14,000 years the Azores were a part of a much larger land mass, perhaps Atlantis, which has since subsided.

Ignatius Donnelly's claim that the Azores are "undoubtedly the peaks of the mountains of Atlantis" seems plausible in the light of the intense volcanic activity that continues on into the present. In 1808, San Jorge suffered an eruption that quickly built to 3,500 feet and kept on for six days. Three years later, a volcano built a cone from the ocean bed near San Miguel, forming an island 300 feet in elevation above the surface of the sea. It was named Sambrina, but soon subsided under the surface. Earlier eruptions of this magnitude had occurred in the Azores in 1691 and 1720.

Perhaps the most important evidence supporting the hypothesis of a sunken land in the Atlantic comes from an oceanographic expedition during 1947–48. Examination of deep sea cores taken by the Swedish Deep Sea Expedition aboard the *Albatross* revealed a freshwater species of diatoms (algae) in a marine environment *under* marine sediment. The diatoms were found in core samples taken from a depth of nearly 2 miles on the mid-Atlantic Ridge. In one of these cores (number 234), more than sixty species were found in great numbers in a layer *exclusively* composed of freshwater types. This discovery led Swedish paleobotanist R. W. Kolbe to report in *Science* that he had observed a freshwater sediment. Earlier, core 234 had caused another Swedish scientist, geologist Rene Malaise, to theorize that these freshwater diatoms lived where they were found and had their origins in lakes or other freshwater habitats. According to Kolbe, Malaise considered this evidence for the existence of Atlantis in his work, *Atlantis, en Geologisk Verklighet.* The coordinates given by Kolbe for core 234 indicate that it came from the Sierra Leone Rise (an eastward extension of the mid-Atlantic Ridge) 578 miles off the coast of Africa. The shallowest part of the Sierra Leone Rise near this spot is 8,000 feet deep. (In the Piri Re'is map, one of two large islands shown in the Atlantic was located on the now submerged Sierra Leone Rise.) Malaise has written that, from the present terrain of the sea bottom, he has deduced that the mid-Atlantic Ridge was shaped less by undersea currents than by above-water weathering processes.

In February of 1969, far to the west where the Atlantic Ocean meets the Caribbean Sea, another enigma was found on the seafloor. Working on the submarine Aves Ridge, which stretches underwater from the Virgin Islands to Venezuela, a Lamont-Doherty Geological Observatory team from Columbia University dredged up more than a ton of granite rock. Granite had never before been found in the open ocean. The expedition's scientific consensus was that either a lost continent had once risen from these waters or that nature is now building a new continental mass in the area.

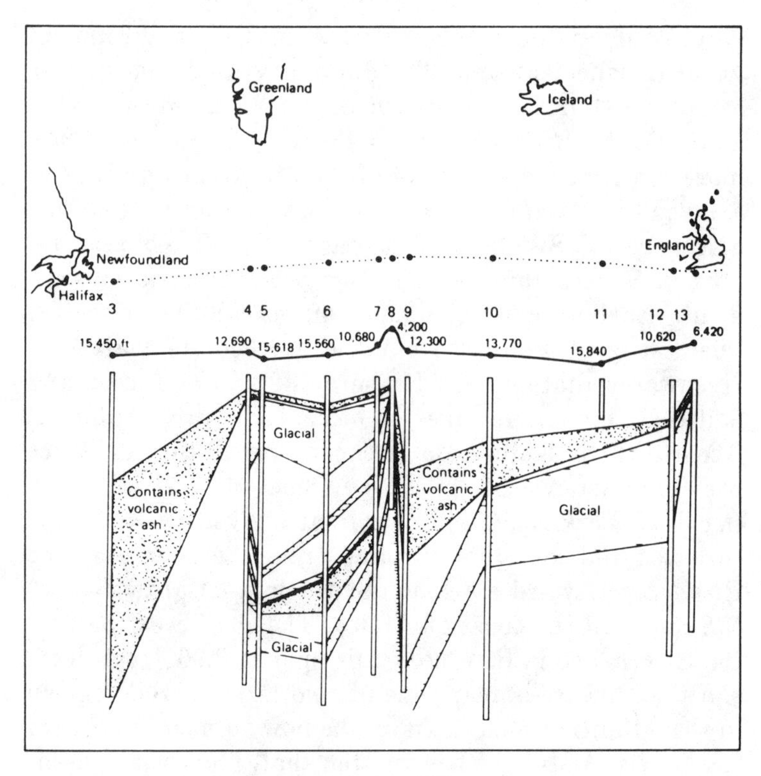

Core holes in the mid-Atlantic Ridge, taken by Dr. C. S. Piggot in 1936. (From *The Life and Death of Planet Earth* by Tom Valentine, Pinnacle Books.)

Further insight into the instability of the seafloor along the mid-Atlantic Ridge comes also from cable work. In 1923, only twenty-five years after the tachylite find, repair operations were conducted on the same site. It was found that during the intervening years—a mere instant of geological time—the seafloor had risen 4,000 feet. Another illustration of the instability of the ridge is to be seen in an event on an extension, the Reykjanes Ridge. From November 1963 to June 1966, 20 miles southwest of Iceland, volcanic activity on this ridge created a new island, Surtsey. This island is now forming permanent vegetation and has since been joined by two other islands. Similar activities may be expected to continue in the Atlantic basin because, as Dr. Maurice Ewing of the Lamont Geological Observatory puts it, the deepest rifts of the Atlantic "form the locus of an oceanic earthquake belt."

The thirteen small Canary Islands, located 50 miles off the coast of the Spanish Sahara, are another area of interest to Atlantologists. Their terrain includes mountains whose elevations reach more than 12,000 feet. The sunny, temperate climate that allows the islanders to grow geraniums, lilies, dahlias, roses, figs, olives, sugarcane, and bananas is similar to that described by Plato for his Atlantis.

Plato's physical description of the island continent (which, remember, perished overnight from volcanic activity), cited black, white, and red rocks. These colors are common in the volcanic rocks of both the Azores and the Canaries. Plato's claim of a temperate climate that allowed the production of inexhaustible fruit supplies also fits both the Canaries and the Azores. The latter island group also has the hot and cold springs Plato mentioned, while the Canary island of Tenerife has a candidate for Plato's great mountain rising from the central plain of Atlantis, the 12,198-foot Mount Teide. In view of Plato's stress on lofty mountains, I found it interesting that this mountain in the Canaries rises 24,798 feet above the Abyssal Plain.

Several unexplained biological phenomena are found in the Canaries. Two thirds of the butterflies there and in the Azores are species also found in Europe, and one fifth are common in America. A mollusk, *Oleacinidia,* whose habitat is Central America, the Antilles, and Portugal, is also found in the Canaries and the Azores. An ancient land bridge such as Atlantis would have made these species' presence easier to understand. A small blind crustacean, *Munidopsis polymorpha,* exists nowhere but in a black saltwater tidal pool near Cueva de los Verdes on Lanzarote in the Canaries. A species related to these blind crustaceans, not itself blind, "lives at what may be the submarine exit of this Atlantic pond," says Charles Berlitz. Has this tiny crustacean been biologically marooned by the sinking of Atlantis? These and many other facts added to my growing conviction that if Plato's story were true, the location was certainly in the Atlantic.

During the last Ice Age the level of the Atlantic Ocean dropped over 400 feet, then began to rise again about 15,000 years ago as the ice melted. Furthermore, the rise was not at a uniform rate. As reported in Professor Emiliani's work, a dramatic rise in water level took place about 11,500 years ago. Such changes in the sea level may have drastically affected the extent of dry land in the Atlantic Basin. Jean Albert Foëx, author of *Histoire sous-marine des hommes* (The Underwater History of Man) feels that the low water after the ice ages (450–500 feet below present levels) would have expanded the continents to the limits of the continental shelves. The Bahamas would have encompassed the Great and Little Bahama Banks, and the Azores and Canaries would also have been enlarged. More recent work refines the oceanographic data from which Foëx worked, but his theory may have relevance for the colonial sites of Atlantis after its destruction. They would have lost their land area much more slowly, the inhabitants gradually being forced to migrate elsewhere. In 1968 John D. Milliman and K. O. Emery of the Woods Hole Oceanographic Institution, reporting in *Science,* established a lowest sea level of 425 feet below the present level, about 13,000 B.C. From this date there was a rapid rise until about 5000 B.C. Other evidence indicates that the sea level changes were accompanied by tectonic violence in the earth's crust, including volcanic eruptions. Professor Emiliani's article in *Science* identified a rapid melting of glaciers at about 9600 B.C. as the cause of flooding for at least ten years.

Emiliani's article confirmed my own growing feeling that the earth's records held clues to a more spectacular past than we have conceived, even during the limited time of human existence on the planet. But what was the cause of all this instability? Were these upheavals isolated occurrences, or could they be related to particular phenomena?

Continuing to study current developments in the earth sciences, I learned that our planet is affected by complicated electrical and magnetic fields that shield life on earth from cosmic radiation and permit the continuation of life as we know it. In the past, however, these fields have gone through at least 170 reversals, each time causing the earth's field to go through a null (or period of zero strength), a condition that would minimize the shielding effect—or, in other words, allow increased cosmic-particle penetration, or radiation, into the earth's atmosphere. Some scientists are beginning to suspect that the end of each geological period, now calculated by the demise of certain species of animals, may also coincide with these field reversals. For example, dinosaurs may not have become extinct only because of a simple change of climate that would have affected their food supply, but may instead have been irradiated into extinction. Perhaps, I theorized, humankind itself had been decimated a number of times, triggering cultural regression or a need to begin all over again. If so, this would help to explain how a technically advanced culture such as Atlantis disappeared perhaps during an upheaval of global proportions.

The mechanism that produces earthquakes and volcanism on the earth's surface is clearly plate motion. But what is the timing mechanism or trigger? Several have been proposed. Don Anderson, head of the seismology lab at the California Institute of Technology, has suggested slight changes in the earth's spin rate and also a thermodynamic episode in the mantle—a heating up of the earth's burners, as the geologist Dr. Jeffrey Goodman would put it. Others have suggested tidal effects from the moon. Plate motion builds up strain; many possible triggers have been identified.

Some years before the present work and these recent suggestions, the correlations between polarity reversals, volcano activity, and mass extinctions had led me to the idea that some change in the earth's geomagnetic field somehow triggered seismic and volcanic activity to release pressures built up by plate motion.

Just before a 1964 quake with its epicenter near Anchorage, Alaska (Richter 8.5), petroleum engineers noted an unusual change in the earth's magnetic field. Also, in 1973, a change in the ionosphere, which disabled the Navy's Omega navigation system (operating at 10–12 hertz) for 45 minutes, was accompanied by an earthquake in the Hawaiian Islands with its epicenter at Hilo.

Both these events led me to a conception of the earth's magnetic field as a magnetic bottle that somehow retards plate motion—until it weakens, as happens during either a short-term or long-term reversal of polarity.

How could the geomagnetic field be changed or change its strength? Even the exact source of the field is disputed, but the usual explanation is that a dynamo mechanism, or motion within the outer portion of the molten core of the earth, generates the field and periodically reverses its polarity.

In addition to these internally caused changes, some

have speculated that a near passage of a comet or asteroid might reverse the polarity. A phenomenon which is easier to track is the relationship between the sun's activity and geomagnetic field intensity. Solar flares appear to correlate with wide swings in field intensity.

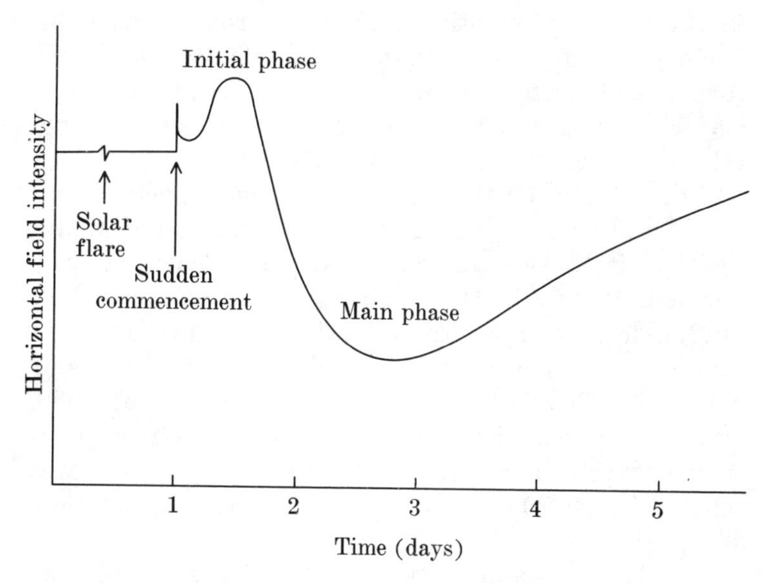

Typical response of geomagnetic field during a magnetic storm induced by solar flares.

When a strong solar flare's energy reaches the earth, it causes heavy currents to flow in the ionosphere (an ionospheric storm) about 50–250 miles above the earth's surface. These heavy currents first induce an *increase* in the geomagnetic field's intensity and then a *decrease.* It is possible that it is this weakening that results in a reduced containment by the field and allows a momentary increase in plate motion—and the resulting seismic and volcanic activity. Gravitational anomalies have also been noted at the times of such activity; perhaps there is also a relationship between the geomagnetic field and the gravitational field.

A series of events after the first edition of the present work suggest the possible validity of the electrodynamic hypothesis. On May 18, 1980, Mount Saint Helens had its first major eruption and two strong quakes hit northern California. At that time, seismologists in California denied any connection. Eighty hours earlier, a major solar flare had been observed. Its effects were not felt on the earth's surface until the morning of May 18—the time when the seismic and volcanic activity hit the West Coast, including the major eruption of Mount Saint Helens. This correlation was revealed to me in a telephone call to the High Altitude Research Laboratory in Boulder, Colorado, the following day. As a result, I was led to compare major seismic and volcanic events with solar-flare activity for the entire year of 1980. During that year there appeared to be a strong correlation between increases or decreases in solar flares (change of state) *and* volcanic and seismic events globally.

The electrodynamic hypothesis that has emerged is as follows. First of all, conditions for seismic or volcanic activity are established by plate motion, itself the end result of convection currents from the earth's interior nuclear heat engine. Once these pressures reach the critical level, various triggers may release them. The release may take the form of a volcanic eruption or an earthquake. In addition to proposed triggers such as changes in the spin rate, increased activity within the mantel, or lunar tidal forces, solar-induced geomagnetic storms may (through a weakening of the geomagnetic-field intensity) allow plate motion to accelerate, thus producing seismic or volcanic activity. Again, in this hypothesis, the geomagnetic field is assumed to function like a mag-

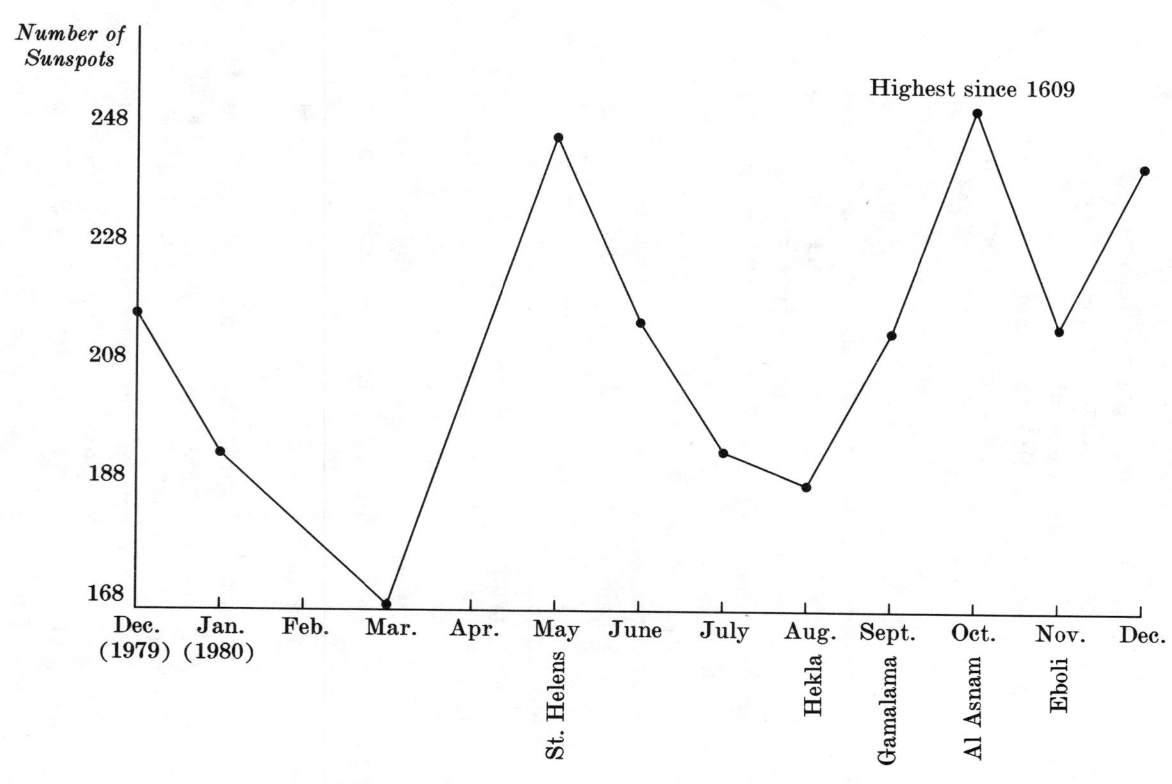

Correlation between 1980 sunspots and seismic-volcanic activity. (Increases in sunspot activity parallel solar flare increases.)

netic bottle that usually restrains plate motion. It is also recognized that when plate motion has built pressures to a critical level, then any one of the abovementioned triggers could operate.

Interestingly, in 1981, Marsha Adams of the Stanford Research Institute reported a connection between solar-flare activities and earthquakes. Other scientists debunked her hypothesis, saying she had no model to show a relationship between these phenomena. The above-described electrodynamic hypothesis provides the model these scientists claimed Dr. Adams could not supply. Clearly she was on the right track; just as clearly her observations generated "anomaly anxiety" in the scientific establishment.

If temporary weakening of the earth's magnetic field triggers seismic and volcanic activity, the process of reversing the field (which includes 100–1,000 years of a weak and confused field strength) would certainly step up such phenomena.

Evidence from marine paleontology gives further weight to the theory that geomagnetic-field reversals are responsible for the mutation or extinction of various life forms on earth. It seems that when the geomagnetic field is weakened during a reversal, the bombardment of solar radiation may diminish the ozone shield that protects life on earth from the intense ultraviolet radiation from space.

A series of events that occurred about 690,000 years ago will illustrate the possibilities. This was the time of the Brunhes geomagnetic reversal. Possibly the weakened field and subsequent solar and cosmic radiation caused excessive nitric oxide to form and, in turn, caused the destruction of part of the ozone shield. Several forms of microscopic sea life became extinct at this time, according to George C. Reid and I. S. A. Isaksen.

To make this time in the earth's history—the era of

Peking man and Java man—more dramatic, a glass hailstorm from outer space showered all across the Indian Ocean. These glass particles, or tektites, included the known habitations of humans in their path which extended from the South China Sea through Indonesia to Australia and almost to the tip of South Africa.

Another significant reversal of the earth's magnetic field took place about 40,000 B.P. (before the present). This reversal was detected at Lake Mungo in Australia. Often the evidence of reversals is imprinted in seafloor rocks as they cool down from a spreading rift zone such as the mid-Atlantic Ridge. The Mungo reversal, by the mechanism identified above, apparently coincided with the eclipse of the Neanderthals and the rise of the Cro-Magnons, who have been dated as early as 38,000 B.P. This connection between human evolution and geomagnetic field reversals was proposed by John S. Kopper and Stavros Papamarinopoulos in the *Journal of Field Archaeology* in 1978.

Another reversal observed at Lake Mungo at 30,000 B.P. may possibly relate more directly to the Atlantis problem. Some Atlantologists have connected the Cro-Magnons with Atlantis, and the Edgar Cayce readings record a migration 30,000 years ago from Atlantis to the Yucatan because of violent seismic activity on Atlantis. The timing of this migration is consistent with the implications of the electrodynamic hypothesis described earlier.

I have no doubt that volcanic activity was considerable in that era. The instability of the earth's crust is evident in the violent explosion of Santorini in the Aegean 5,000 years later or about 25,000 years B.P. The debris of this mighty explosion was deposited across the floor of the Mediterranean almost to Italy.

Finding in all this data strong suggestions of a large-scale drama of prehistory, I then looked at the evidence for a catastrophe at the time claimed for the final destruction of Atlantis. Plato's account gives us the date of about 11,500 years ago for the final destruction of the island continent; 12,350 years ago (dated geologically), the earth's fields again reversed in what is called the "Gothenburg magnetic flip." Checking Edgar Cayce's material, I found that his date for the final destruction of Atlantis was 12,700 B.P.—extremely close to the time of the magnetic flip.

On Plato's side, Professor Emiliani has established the time of an ancient flood at about 11,600 B.P., dated from the growth rate of *Foraminifera,* a marine microorganism found in ocean sediments. The carbon-14 dating method also gives us the date of 11,240 B.P. for the arrival of a big-game-hunting people in Clovis, New Mexico. Where did they come from? Incidentally, 11,000 B.P. is the conventional date for the end of the last Ice Age. All these dates can be included within 1,700 years—quite a narrow time span considering our dating procedures, the different sources of these dates, and the length of time from our own era. Perhaps there were two dramatic earth events about 1,000 years apart, say 12,500 and 11,500 B.P.; but whatever the precise sequence, prehistoric peoples apparently witnessed a mighty drama played out in their environment.

9

AN EXODUS FROM ATLANTIS?

More at ease with my commitment to fieldwork at Bimini, I decided to look more carefully into anthropology for new light on Atlantis as a source of world cultures. Again, contemporary thinking about ancient migrations was found to be undergoing reexamination.

The conventional theory of east-west migration, represented in Jacob Bronowski's contemporary work, *The Ascent of Man,* claims that man's ascent began in the Fertile Crescent of the Middle East, then followed a linear rise from a nomadic hunting culture, to farmer, villager, and then city dweller in such ancient civilized centers as Ur, Jericho, and Ugarit. From here, humans spread out to the west. However, this view is being challenged by recent developments in radiocarbon dating. In his book, *Before Civilization: The Radiocarbon Revolution and Prehistoric Europe,* Colin Renfrew argues that the correction of carbon-14 dates by comparison with ancient tree rings establishes that Europe's megalithic structures (Stonehenge, for example) are *earlier* than and *independent* of Egypt and Mycenae.

Bronowski's theory also ignores the incredible intellectual development of the Egyptians during the third millenium B.C., an achievement whose true extent is just beginning to be recognized after centuries of investigation into the pyramid of Cheops at Giza. While Stonehenge indicates a surprising awareness of astronomy, the Great Pyramid of Cheops testifies to the incredibly sophisticated Egyptian science of the third millenium B.C. This could imply the survival of knowledge from an earlier civilization whose emigrants shared their knowledge with the Egyptians. A recent summary of many scholars' work on this structure is found in Peter Tompkins' book, *Secrets of the Great Pyramid.* The pyramid, to begin with, has a north–south orientation which is a standard for modern compasses. It is also a precisely located marker that was the basis of ancient geography, as well as a celestial observatory. Even the value of pi is incorporated in the structure, as are the sacred triangles attributed to Pythagoras, the 3-4-5 and $2\text{-}\sqrt{5}\text{-}3$ ($a^2 + b^2 = c^2$) triangles, which Plato, in his *Timaeus,* claims to be the basis of the universe. Further, the Great Pyramid reflects knowledge of the precise circumference of the earth, the length of the solar year to several decimal places, and so on. According to Tompkins, its designers apparently knew "the mean length of the earth's orbit around the sun, the specific density of the planet, the 26,000-year cycle of the equinoxes, the acceleration of gravity and the speed of light."

Although the Great Pyramid also seems to be more evolved in its mathematics and astronomy than Sumer or Babylonia, doubtless the Egyptian civilization will eventually be recognized as *earlier* than the cultures of the Fertile Crescent. The present conventional date for the beginning of construction of the Great Pyramid is 2644 B.C., but the amazing structure certainly hints at a more sophisticated culture than any offered by that part of the world at that date. Tompkins feels that further discoveries "may open the door to a whole new civilization of the past, and a much longer history of man than has heretofore been credited." I have shared this feeling for a number of years.

I recalled that Lewis Spence also advanced a pro-Atlantis argument for a migratory pattern from west to east, supported by evidence that Cro-Magnon remains, originally found in Western Europe, were excavated as far away as Brazil. How did they get there? If Atlantis was the Cro-Magnons' place of origin and he fled during one of the migrations, it would not be surprising to find their skeletal evidence on both sides of the Atlantic. Once again, while parallel culture traits are not conclusive of themselves, a recent discovery adds to the overall probability of an early transatlantic diffusion: flint points in Sandia, New Mexico, were dated to about 25,000 B.P. The Solutrean flint, a very similar type of flint point with the same dating, was found across the Atlantic in Morocco and France. In addition to this possible indication of Cro-Magnon presence in the New and Old Worlds, the dating is the same as Spence's claimed date for one of the Atlantean destructions.

The theory that humans came to the New World only from Asia, via the Bering Straits, is also being challenged by new geological findings. Growing evidence indicates that the land bridge between the continents, originally covered by water 5.5 million years ago, was later open, 100,000 and 70,000 years ago. Early human presence in the New World has been dated at least as far back as 50,000 years ago. Dr. Alexander von

Wuthenau, an authority on pre-Columbian art, has discovered that artifacts of both Central and South America portray all racial types. He contends that this supports the theory of migration by all groups to these areas. In short, the evidence is beginning to suggest a *very* ancient and complex migration pattern, worldwide in scope.

And what of the mysterious stone sites that suggest the flow of astronomical knowledge from Central America north to the American Indians?

Sites from Wyoming, known as the Amerind Stonehenge, up through to the Medicine Wheels of Canada, have recently been linked with England's Stonehenge as examples of astronomically oriented centers. What was the origin of this knowledge? Some Mesoamerican experts feel that Mesoamericans originally came from as far south as the Amazon Basin, and I am convinced that Brazil holds important evidence for the student of Atlantis.

In 1986, in a cave about 900 miles north of Rio de Janeiro, a Brazilian archaeologist, Maria Beltrão, found a primitive tool with animal bones that were later dated by uranium-series tests to between 350,000 and 200,000 years ago. This and other recent Brazilian discoveries suggest that early humans reached the New World much earlier than the conventional time frame of 20,000–12,000 years ago.

Another challenge to the theory of migration from the Bering Straits comes from hematology. Although East Asian blood types B and AB comprise 30–60 percent of the population in that area, the American Indian has a very low incidence (0–2 percent) of these types. The Basques of the Pyrenees, considered by many as prime candidates for an Atlantean colony, display a similar rarity of B-type blood. These low proportions cast doubt on an exclusively Asian migration.

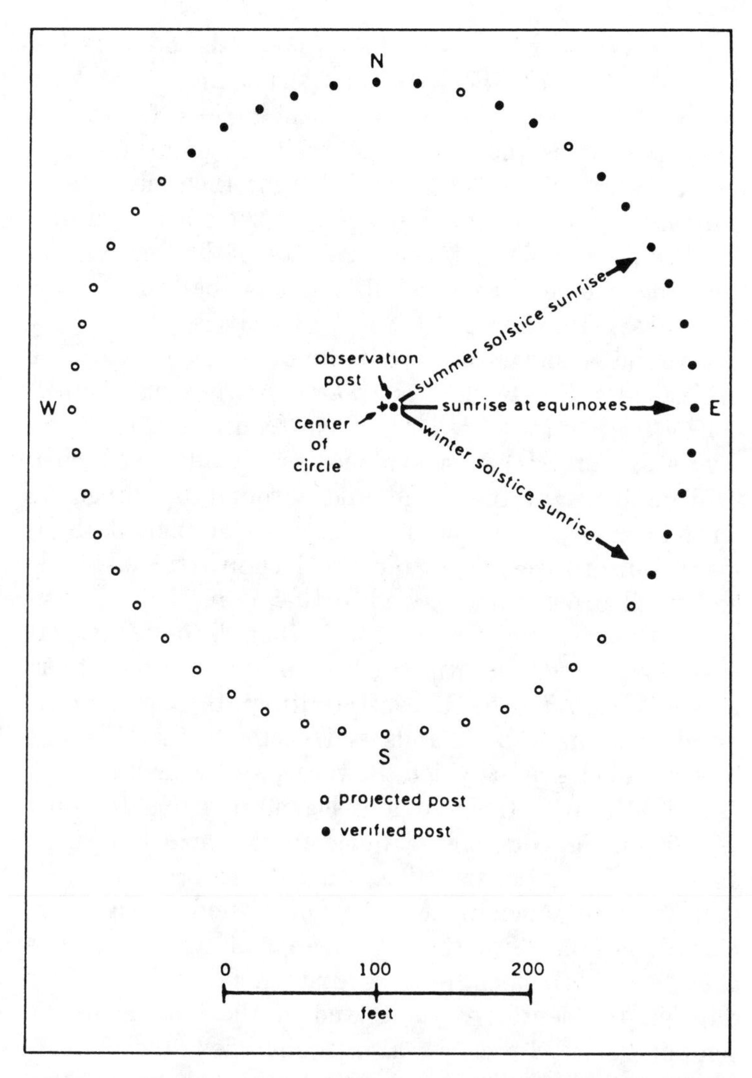

Plan of the American Woodhenge, Cahokia, Illinois. (British Crown Copyright. Reprinted by permission of the Controller of Her Brittanic Majesty's Stationary Office.)

The Guanche culture of the Canary Islands, possibly in existence since the Neolithic period, may also be a link with Atlantis. Perhaps the most curious fact about the Guanches is that they possessed a legend of a universal, catastrophic flood and, believing themselves to be the only survivors, were surprised at their first visitors. Furthermore, they gave no evidence of having been a seagoing people. The possibility exists that they originally crossed a land bridge to the Canaries.

I found a substantial contemporary account of the Guanches in *The Silent Past: The Mysterious and Forgotten Cultures of the World* (English translation 1962) by Ivan Lissner. He gave evidence of an advanced Neolithic culture in the large underground buildings on Grand Canary, sites that are suggestive of ancient Mediterranean cultures. The Guanches lived in artificial caves, had small circular houses and fortifications. Their physical type was that of Cro-Magnons, though they may perhaps have later intermarried with the Berbers from North Africa. In the fifteenth century, the Spanish arrived to occupy the islands and decimate the Guanche culture, evidence of which, however, still remains.

In addition to their curious legend of being the only survivors, the Guanches claimed to have had ten kings, as in Plato's Atlantis. (The ten rulers of Atlantis, by dictate of the ancient authority of Poseidon, ruled according to the laws the first men had inscribed on a column of orichalcum—a reddish metal unknown today—in the center of the island at the temple of Poseidon.) The Guanches also mummified their dead, although not in the same fashion as the Egyptians. Present-day Canary islanders still wrestle in a manner similar to that depicted on Egyptian bas-reliefs of the Middle Kingdom. The Guanches also practiced sun worship; that is, like the Egyptians, Mayas, and Incas, they had a solar cult.

The question of whether they had a written language has not been resolved. Thus far, the strange symbols on the rocks of La Palma and Hierro archaeologists have not accepted as a form of writing. In Robert Charroux's *Forgotten Worlds,* he and his wife, Yvette, describe "writing" they found on a basaltic outcropping in a valley at the center of Grand Canary: "We recognized and photographed designs similar to those of the Celts in Brittany: spirals, circles, serpents, stylized persons, and a remarkable "sorcerer" that is an exact replica of the one in the Villar cave in the Dordogne, France. Still more important, we took pictures of genuine writing and submitted them to Abbé Hirigoyen and the magazine *Découvertes* for expert examination. The writing is unquestionably composed of letters, some of them resembling our letters V, N, S, T, and I."

Where did this language come from? Consider, too, the mysterious decline of the Mayas of Mexico who, despite a highly evolved society, inexplicably disappeared about A.D. 900. These people have long been of interest to Atlantologists. Cultural parallels between the Mayas and the Egyptians seem to support a theory of migration from a land mass in the center of the Atlantic to both the east and west. I later learned that the Mayan culture had been pushed back in time 1,700 years, to the third millenium B.C.—thus the Mayas paralleled the Egyptians in time! Modern research into the Mayas has only added to our knowledge of their achievements in mathematics and astronomy, especially puzzling for a Stone Age culture. The Mayan calendar suggests more precision in handling time than is to be found elsewhere in the West until the beginning of modern astronomy.

Both Mayan and Egyptian cultures used pyramids for astronomical observations (the Mayas at least as early as 500 B.C., in Monte Alto, Guatemala). Among the achievements of ancient Egyptian science was the observational recognition of such subtleties as the semidiameter of the sun and the 26,000-year precession (the slow movement west of the equinoxes), both probably as early as the third millenium B.C.

While certainly not at this level, the Mayas nevertheless reveal a considerable astronomical awareness, particularly if we regard them as a Stone Age culture. Their calendar, created 1,000 years earlier than the Gregorian version, was, as adjusted, closer to the solar year. In *Stonehenge Decoded,* Gerald S. Hawkins explains that this exactness was achieved through observational corrections of the 365-day calendar. Mayan pyramid observations included the declination of the sun at summer and winter solstices, the equinoxes, winter and summer moon declinations, and the patterns of Jupiter and Venus. As Hawkins points out, this care gave them the exact period of the moon: 29.53 days. (It has occurred to me that the need for this observational correction may explain one of the uses of the Long Count of days and the two yearly calendars, the solar [365 days] and the sacred [260 days], which two calendars coincided every fifty-two years [or 18,980 days in the Long Count].) Another Mayan achievement is the unusual numbering system based on twenty instead of the usual ten. Adding to the transatlantic coincidences, the system of twenty is also found among the Basques, as well as in other cultural "islands" such as the Celts, the Hamites, and the Ainus.

As I began to recognize the implications of the diverse theories, scientific findings, and psychic information, I was ready to consider the most fantastic data I had yet been given—that of a psychic who had joined us on the '75 expedition. Her incredible tale could now be evaluated in the light of all the evidence I had accumulated from other sources. Perhaps I could then determine my own theory of the origins of the Bimini Road site.

PART III

THE PLEIADES THEORY

10

PSYCHIC ARCHAEOLOGY

During the course of the Poseidia '75 expedition, as the probability of Bimini as an Atlantean site loomed larger, I decided to call in help from another quarter.

With some satisfaction I had learned that other researchers into prehistory are beginning to admit to using psychic data in their fields, partly due to the accuracy of Edgar Cayce. In 1937, during the course of a past-life reading, Cayce spoke of the subject's association with Jesus in a school run by the Essenes, located on the northwest shore of the Dead Sea. The first of the ancient Essene manuscripts, later known as the Dead Sea Scrolls, was not to be discovered until 1948. The Essene community itself was not discovered in archaeological digging at Qumran until 1949. Cayce also implied a close working relationship between Jesus and the Essenes, which is still being investigated by conventional historical and religious circles. Had Cayce's work been better known in 1937, his lead to the Qumran site might have been heralded as a first in intuitive archaeology. As it was, Cayce died in 1945 without hearing of the discovery he himself had pinpointed.

Today, a number of sensitives are offering information on the various functions of ancient sites as yet unconfirmed by standard archaeological procedures. By working with psychics of proven ability, we can construct testable hypotheses, both about ancient sites and the mental development of their builders. Paranormal channels played an important role in our friend Dr. Valentine's discovery of the Loltun cavern in the Yucatan.

At the Mexico City meeting of the American Anthropological Association in 1974 at least two scientists gave papers describing their use of psychics to conduct archaeological investigations. One of them, Dr. J. Norman Emerson, of the University of Toronto, described his system of evaluating eleven different psychics who were asked to identify the age and maker of certain Iroquois Indian artifacts. The most important paper on the subject of paranormal data in archaeology was probably Jeffrey D. Goodman's "Psychic Archaeology: Methodology and Empirical Evidence." In it, Dr. Goodman details his use of psychic information from 1971 to 1973 in furthering his work on a site near Flagstaff, Arizona. He had asked a psychic (by profession an aerospace engineer) to locate a "new" early site for him. After being denied a permit because of his unorthodox approach, Goodman was given a compromise permit to dig in an area where experienced archaeologists had said that artifacts would be "very unlikely." He succeeded in locating the first known deep excavation in the area.

Goodman credits his psychic assistant with: "(1) cor-

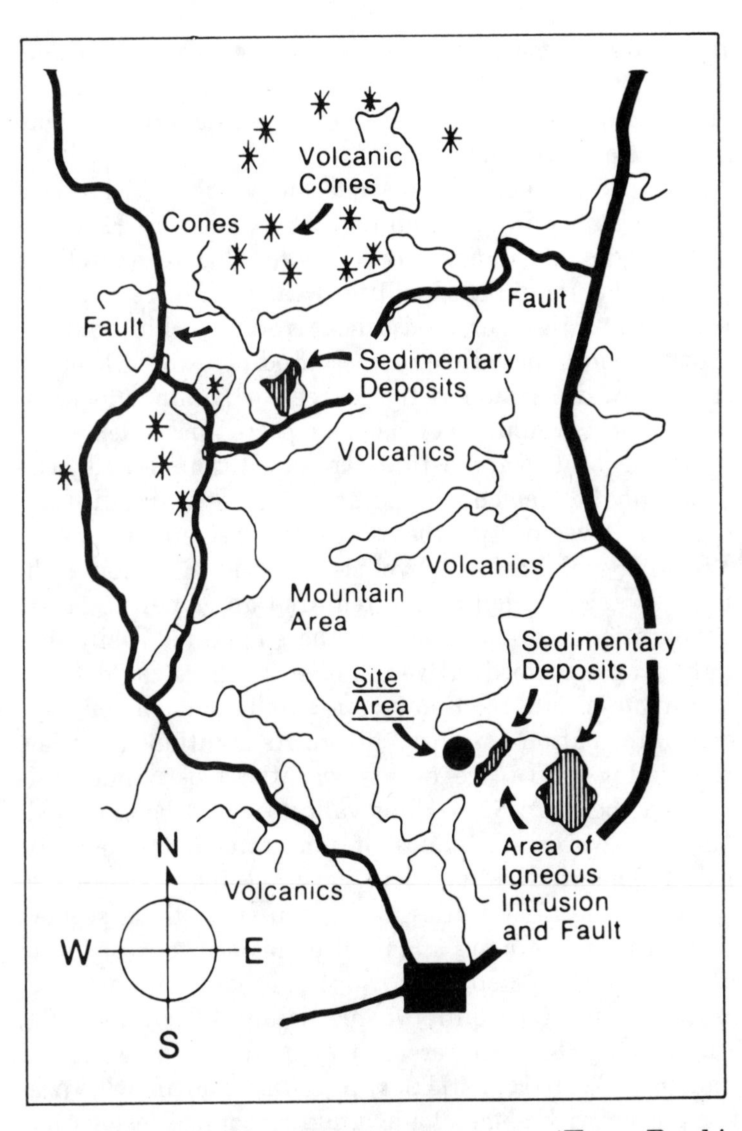

Map showing Goodman's archaeological site. (From *Psychic Archaeology* by Dr. Jeffrey Goodman, G. P. Putnam's & Sons. Reprinted by permission of The Berkeley Publishing Group.)

rectly locating a site inconsistent with the traditional model; (2) "correctly predicting the key geologic and archaeologic contents; (3) "discovering a new Early Man site of major importance."

I suspect that intuitive archaeology will gain increasing acceptance in the coming years, thanks to Emerson and Goodman's pioneering attitude. Our own findings during the Poseidia expeditions earned my respect for the possibilities for assistance from psychic sources. But the most incredible of these findings would lead me into a new dimension of the search for Bimini's lost history—one I would never have suspected possible.

For several years before the expedition, I had been aware of the tremendous potential for psychic information to develop hypotheses about prehistoric events. Working with various psychics, I found that their techniques varied widely: some required an elaborate ritual preparation and trance state; others operated easily in a light trance or meditative state; still others could function from a fully conscious state and comment on psychic information as it was given. As mentioned earlier, the timing of future events was the most undependable—at best, an 80 percent validity could be expected. In other words, eight out of ten statements would be valid. Some individuals had very specific responses to earlier physical activities at a given site; others seemed to be attuned to more spiritual or metaphysical aspects of the past. Some spoke in riddles; others gave a coherent, sequential account of prehistory. Then in 1974, while doing Kirlian research at Lamar University, I had met Carol Huffstickler, a young woman who presented me with some challenging concepts. Besides her work with auras, Carol had worked as a psychic subject in telepathic dreaming experiments at the Maimonides Dream Laboratory in Brooklyn, under Drs. Stanley Krippner and Montagu Ullman.

When I talked with Carol about Poseidia '74, she asked me whether I would like a psychic reading on Bimini. When I said that any information would be helpful, she clairvoyantly tuned in to the past. Bimini, she said, had experienced a cataclysmic land change in 6031 B.C., with residual effects lasting for 200 years. The culture we know as Atlantis continued until 4021 B.C., with some of the population still in the Bimini area, though most of it had emigrated to the British Isles and Peru. As she watched the earth changes, she saw a slipping of its crust, moving Bimini southeast to its present position. Although the timing wasn't entirely clear, the Bimini site itself had gone beneath the sea at the time of the cataclysm in 6031 B.C. Her date proved consistent with an ancient beach line now 50 feet under the sea, which has been dated by geologists to about 6000 B.C.

Before our departure for the 1975 Poseidia expedition, I had asked Carol about the hazards—if any—of the Bermuda Triangle. She warned us not to dive or fly during the summer solstice. As if to confirm her warning, the first day of the solstice blew in with squalls with cloud tops up to 50,000 feet high, 40-knot winds, green skies, and other violent signs of summer tropical storms. This wild weather continued for the entire solstice period of June 18–23. Even between storms, the sea bottom was too disturbed to allow us decent diving visibility. But after the end of the solstice on June 24, the weather improved dramatically.

Now as the late sun silhouetted the restless palms, one of Chalk's twin-engine Mallard seaplanes waddled up the ramp on Bimini, its engines roaring. Fourteen passengers struggled clear of the door, among them Carol, obviously tired from her day's flight from Houston. On top of the ordinary wear and tear of travel, she had experienced the cultural shock of the transition from Houston's space age airport to Chalk's cramped aircraft and Bimini's tropical setting. Once settled in, however, Carol proved of help on several occasions during the 1975 expedition.

While still in Houston, using a chart of Bimini to focus on, Carol had identified the site of a suspected structure on the banks of South Bimini. When C. W. Conn flew me over the site at the beginning of this expedition, I was delighted to see a definite pattern on the sea bottom which, in outline, resembled a Mayan temple. I photographed it as we passed overhead, but we were unable to relocate it on subsequent flights. This was the first of many turtle grass patterns which seduced researchers.

After spending several hours photographing and measuring a shipwreck for the Bahamian Antiquities Institute, I discarded the popular theory that it was Phoenician. Its visible technology suggested that it was hardly more than a century old. Later that day, Carol did a reading on the wreck, and described it as the *Gloria Victoria,* hinting at an English origin. It was a three-masted sailing vessel, she reported, bound with a cargo from an east coast port for a Georgia plantation in about 1840. During a storm the ship was blown off course and sank. This reinforced my own conclusions from on-site observations—that the vessel belonged to the last century and had either foundered or was sunk by pirates. The marble blocks in its cargo seemed to have been shaped with the English foot of 12 inches in mind.

A subsequent dive by the BAI uncovered a safe, bearing a bronze identification plate with the name of the model, the Improved Defiance Salamander, and of the manufacturer, R. M. Patrick, a New York firm. We checked this lead thoroughly and finally determined that the safe was dated sometime after 1843. Carol's timing appeared to have been validated.

Carol Huffstickler arrives on Bimini.

On the afternoon of Carol Huffstickler's first psychic reading on Bimini, its drama was foreshadowed by a heavy thunderstorm crashing over the island. We were all curious to learn what she might be able to reveal about the site. No one present, however, was prepared for the incredible story about to unfold.

When Carol's ritual preparation was complete, I asked her to comment on our work priorities. My own feeling was that the Road site was our most important objective. Carol confirmed this and, as I had half expected, said that it was *not* a road. What confused our present evaluation, she continued, was that many of the stones now resting flat on the bottom had once stood vertically.

The site originally included "entrance ways and anterooms and rooms behind rooms, so that it was much like a labyrinth." The labyrinth function would be evident, however, only if a number of stones were returned to their vertical positions. Carol connected the labyrinth function to the site's sacred nature: Labyrinths are historically known as representations (either as symbolic diagrams or as actual structures) of the intricately winding path of initiation to wisdom. When the initiate arrives at the center, he or she experiences a total change of consciousness and experiences ultimate truth.

So Bimini, as well as being a sacred site, might also have been a place of spiritual initiation, a holy place containing special energies! One way I could test this would be to compare the experience of several psychics at particular locations on Bimini, and see whether they independently picked up strong energy at the same places. If so—and if these locations corresponded to significant points in our underwater survey of the site—we would have positive input toward establishing Bimini as a sacred site. (Incidentally, when sensitives visit any of the known ancient holy places on this planet such as Chartres Cathedral in France, Machu Picchu in Peru, or Teotihuacan in Mexico, they have been known to experience heightened consciousness in response to the energies of the site or structure. Some of the effects may be a tingling over the body, or even the stiffening of the hairs on the arms. The intensity and quality of the reaction can vary considerably according to the psychic's attunement and the energies of the site.)

Carol then proceeded to tell us more about the Road site. Apparently, what remained was only one-fourth to

Bronze plate from the shipwreck.

one-third of the original site. Its present appearance was the result of thousands of years beneath the sea, and also of an earthquake that had displaced the original buildings toward the sea.

My next question for Carol was: "Which culture originally constructed the site?" Her surprising answer was that the builders were the same people who had also constructed the geometric lines on the plains of Nazca in Peru! Good Lord, I thought, when I heard this, what a statement to try to verify or deny! "They were a very dark-skinned people," she continued, "but it is not your Indians—probably the Atlanteans. I feel that. I can't get the word but it is the exact thing [culture] as the landing strip thing [the Nazca lines] because I keep seeing the same thing over and over. I see that [Nazca] and then this [Bimini]."

Our next surprise came when I asked her for an approximate date. I had meant the date of construction, but Carol got two dates: "There is some confusion with this—there are two things I get, thirty thousand and ten thousand."

Again and again during her first reading, the idea came through that the Bimini site was extremely old. Carol's first reading dated the Bimini site prior to Stonehenge, stating that Bimini was contemporary with the Atlantean culture and that it had been destroyed basically for the same reasons that Atlantis had perished: misuse of sexual energy and black magic.

Her next reading brought new revelations. Immediately the Bimini site was described as contemporary with the Magdalenian culture in Europe which we now date between 15,000 B.C. and 7500 B.C. Except, Carol informed me, these dates were much too recent! She also predicted further improvements in dating methods, then added that such improvements would place the construction of Stonehenge at 16,000 B.C. instead of 2775 B.C., as now believed! Once again, she said that Bimini was constructed before Stonehenge: "Stonehenge was built after the deluge, the Bimini Road was built before!"

She identified the Bimini culture as related, although prior, to the Minoan culture. She then got the number fifteen. Evidently this was 15,000 years ago, and seemed to be associated with a moving of the capital city away from Bimini and the abandonment of the original site to a more primitive culture. She said of the Bimini site,

"It is very, very old." She also added that inscriptions from the later culture might be found that would confuse the identification of the original culture. In seeking the site's origins, Carol got the impression of "the seeding of the planet by . . . cosmic races, ones that travel from place to place and seed different planets, teaching a religion suitable to the evolution of those already present on each planet. This was happening in several places at one time . . . but it didn't work very well because there was still a great deal of resistance from [beings] getting involved in matter."

As the full implications of this first reading swept over me, I experienced mixed feelings. On the one hand, I was very excited over the suggested parallels between the reading and the ancient accounts of human origins through the descent of spirit into matter. But extraterrestrial visits were a phenomenon I had not anticipated in connection with Atlantis. And some contemporary writers had treated the topic in such a lurid, mechanistic, and dubious fashion that upgrading their speculations seemed impossible. I began to look for a common denominator between the stories of Atlantis (and souls descending into matter only to be trapped) and extraterrestrial visits. Could these ultimately be reduced to different facets of now-forgotten ancient events on the planet? Did myths contain any usable elements which could be applied to such a reconstruction? Finally, would archaeological discoveries ultimately confirm such theorizing? Fortunately during the day, the physical discipline of diving subdued the wild scenario that was claiming most of my waking moments. In the evening, though, I once again became aware of the tiger to whose tail I now clung.

Then, in Carol's next reading, came the bombshell—although its explosion was delayed because really stunning ideas do not come through all at once. "I keep getting the word Pleiades," Carol said. "I'm going to say it because it keeps coming through. . . . It keeps coming through with the information on the Bimini Road site." At first she thought it referred only to a significant astronomical orientation. Carol brought forth a story of benevolent beings, more highly evolved than humankind today, who visited the earth around 28,000 B.C. from the Pleiades star cluster.

Four hundred light years from our sun and about 600 million years of age, this group of nearly 300 known stars, within the constellation Taurus, travel through space together at a high velocity. In the fall, they rise in the east at evening; in spring, they move west and appear after sunset.

Nine of the stars can be seen without a telescope; most visible are seven, sometimes called the Seven Sisters. The actual names of the cluster, and its brightest stars, are taken from Greek mythology. The Pleiades are associated with Atlantis because *Atlantis* means "pertaining to Atlas," and the Pleiades were the seven daughters of Atlas and Pleione: Maia, the firstborn (and originally the brightest star visible), Asterope, Taygeta, Alcyone, Celoeno, Electra, and Merope.

The Pleions, Carol explained, were evidently on a kind of galactic missionary service. These beings from the Pleiades, as the readings described them, must surely have seemed as gods to the earthlings they visited. "The people that came from this area of the Pleiades are not specifically humanoid of structure as our forms are; their structure is more consciously controlled by the mental form, and so the form is mobile according to the thought pattern. When a being reaches this level of consciousness, it is his moral obligation and duty to observe, to guide and to direct the vibrational frequencies of those creatures who are evolving into a similar pattern of a similar direction; and, because of their consciousness, they have a need to do this. This is part of, as you would say, their life's work. Here on this planet our work is to build and become conscious of who we are; their work, since they know who they are, is to assist others in the process, very much as you would call teachers, except these are not specific inner-plane teachers. . . . Part of the reason of their appearance in this region was to allow the evolution process to speed up [the evolution of] the race present here on this planet."

Upon their arrival, the Pleions discovered the capital city of a continent, "a major center of commerce" and a major religious center. "The beings here on this planet were ready for the contact, and they were in this specific region." Carol went on to describe them as having a solar cult, but did not see any evidence of crops, weapons, or tools. "It just doesn't make sense," she said. "This is what I don't understand: if they are going to build something they are going to have to have tools, unless they are using mind power." This became clearer as her reading continued.

"The culture that was here was telepathic, much more than we are at this time. However, because of telepathic communications there was not much need for action. The reason the Pleiades, or as you say, the Pleions, came here was to begin to stimulate the element called function, or action. Much of what was done here was done without physical function, and they could see the pattern of man having to take physical action . . . then later having to take mental action and reduce physical strength. These are, as you would say, expansion and contraction cycles that the species goes through to produce an end product that is higher up than in the beginning but with the same qualities on a different level. There are trillions of planetary systems similar to ours, and this must be understood: that there are, then, those who are always teachers of these systems and allow them and help them to evolve."

The Pleions chose Bimini as one of their locations for the construction of a sacred site that could use the earth's geomagnetic fields, heightened in some unknown way by periodic alignments of the sun and certain stars, to raise human consciousness and to heal. These beings brought a higher consciousness which they inculcated through the labyrinth and other consciousness-raising features of the Bimini site. The site at Bimini, and apparently others elsewhere, contained specific "operational or standing points" where, in a ritual involving thirteen people including three women, "individuals stood and developed certain currents of consciousness." Apparently the critical aspect of consciousness for the Pleions was the restoring of balance of the masculine and feminine energy.

My own understanding of this rather enigmatic area of the readings is that the balance is that of the typical division between activities of the left brain hemisphere (analytical, logical, linear, thinking—"masculine") and the right hemisphere (synthetic, intuitive, feeling—"feminine") as suggested by contemporary brain research. But as I considered the possible meaning behind the masculine-feminine balancing, I recalled Plato's myth of the creation of humankind. The first beings, he said, were originally spherical in form, and possessed powers so threatening to the gods that they were split into male and female halves. Thus, committed to pursuing each other for wholeness' sake, they no longer threatened the gods. Could Plato's myth be an echo of the male-female balance Carol was claiming as one goal of the Pleions' visit to this planet?

Once, explicitly, the readings identified the culture to which the Pleions had come. It was Atlantis. Furthermore, she contrasted our culture today with the ancient Atlanteans': "We are at the other end of the cycle. You see, we have carried function [action] to its limit and now we are having to go back to the telepathic link with function."

Part of the ritual on Bimini involved a symbolic progression of the seasons, apparently based on the idea that the energy at the site was strongest at the equinoxes and solstices. The most important rituals evidently took place at night. This is consistent with customs of many ancient peoples who began their new year at Hallowe'en, when the Pleiades joined the equinoctial point directly overhead in the night sky.

Although anthropologists have assumed that lunar cults preceded the solar in prehistory, Carol's reading attested to solar cult existence even before the time of the lunar cult. Carol stressed that we were looking at the remains of a solar cult on Bimini, and said that the full history for that site was the original solar cult, then a nature cult, a lunar cult, and finally, another solar cult.

In response to my next question, I was told that a sacred geometry, already suspected because of field observations, would be very hard to discover because "there is too much of it missing now." However, the general orientation, northeast, was "very significant."

Then I asked Carol about the original location of the stones. In response, we were told of a whole cycle of earth changes. The megalithic blocks, she said, came from a mountain range to the east of Bimini that is now submerged. Then she described certain geological movements:

"I see the land of Florida and Georgia all under water. I see the mountains . . . in the eastern United States adjoining onto some of the mountains under water. It looks like . . . the mountains were all together.

The ones under water joined the mountains present now in our eastern coastline, and . . . the Smokies. Then it looks like there is a splitting apart and a sinking of some, then a moving back together, then the splitting apart of land masses and these land masses do not submerge—just the splitting apart. These are the cycles I see. Then the current moves through." This sounded like the Gulf Stream and tended to support a seismic origin for the Gulf Stream's present seabed.

Next she said that the megalithic blocks were similar to rocks now found in North Carolina. The source, however, was now under deep water to the east, in the Atlantic, the first ridge to the east after "very deep water" was encountered. No explanation was forthcoming about how these blocks were transported.

As the second day's session concluded, I was told that the major sites in the Bimini area were originally laid out in a geometric pattern approximately equivalent to the angular relationships between the stars of the Pleiades, "a duplication of the star pattern that was found and viewed from this point." Carol then gave a series of angles ranging from 30 to 62 degrees, including 32, 45, and 52 degrees. Later, using the largest drawing of the Pleiades available to me, I constructed a line that passed through what appears to be the central axis of the constellation—Taygeta, Maia, Alcyone, and an unnamed star. I then drew lines radiating from Alcyone to Pleione and Atlas, from Maia to Asterope, Celoeno, and Electra. Then from the central axis I measured the angles. None matched exactly with those given by Carol, but three angular separations were between 28 and 29.5 degrees, two were 60 degrees, two were 71.5 degrees, and two were 42 and 43.5 degrees respectively.

Two of Carol's angles do fit the Bimini site itself, at least according to our initial survey. The central axis of the outer lead is oriented to 45 degrees (magnetic), the inner to 52 degrees. Neither the plate rotation seen in the readings of another psychic, Joan, or the earth changes seen by Carol would likely affect the internal geometry of the site. Therefore, the 7-degree separation between 45 and 52 degrees (providing subsequent field surveys confirm it) is much more meaningful—as a sacred number, as the traditional number of Pleiades, and as the separation between two of Carol's angles. (In my later readings, I was fascinated to hear that the Notre Dame Cathedrals in France were located in a pattern similar to that of the constellation Virgo.)

Three weeks after the sessions began, we saw Carol off on Chalk's; she was headed back to Houston. As I thought over the material that had emerged in her work on Bimini, I said to myself: "Well, at the very least, it's certainly a colorful hypothesis."

11

GODS FROM THE PLEIADES?

Two years had passed since that summer, and by 1977 I had come to regard the readings in a more serious light. Always aware of the necessity to check out this type of material, I sought to verify the date 28,000 B.C., mentioned consistently by Carol in her readings. Could the site have been constructed that long ago? This would almost predate the conventional age of the Cro-Magnons. This date, however, seems to be the sole element of the readings that conflicts with one of the few known scientific facts about the site: geologist John Gifford dated the seabed beneath the site at 15,000 B.C. by the usually reliable uranium-thorium method. Carol's date of 28,000 B.C. for the Pleion visit seems to conflict with the age of the seabed on which the site rests. Yet, as will later be seen, even Gifford's date is open to question.

I consulted a set of computerized astronomical tables prepared by Professor Anthony Aveni at Colgate University to check the site's 45-degree orientation to see if there were significant astronomical phenomena linked to this orientation at 28,000 B.C. I was surprised to discover that the direction of Aldebaran (the eye of the bull) was favored at that date. I recalled the sacred-bull worship of the Atlanteans and pondered a possible connection. Yet the Pleiades did not rise at the latitude of Bimini until about 18,000 B.C., incompatible with Carol's date of 28,000 B.C.

A possible reason for this discrepancy was suggested by the reading of another psychic I decided to consult on the problem, a long-time, trusted friend whose psychic ability once saved my life. She has been described as "a channel of the highest vibration . . . as close to a pure channel as possible." (In the first edition of this book, I called her Anne.)

I asked her for any additional material—either supportive of Carol's or contradictory. Her own readings brought the following response: "Carol's readings were fundamentally sound. The Pleiades were the source of an outthrust of universal love. The Pleions were beings of radiance and light, emissaries of good will, who aided mankind in times of difficult earth evolutions, although such beings have visited us from time to time from other locations in the galaxy."

Joan, who had been doing archaeological readings on the Bimini site even before the 1974 expedition, found in one of her readings that indeed the site did possess a sacred function, as well as sacred geometry.

I had asked: "Does the Road site contain sacred geometry?" Her reply was: "Yes, assuredly so. The Pleions in like manner were constructing a reminder of the immense undertaking in the world of form . . . the converting of Sons of the Most High into earth nature."

"These [arrow and chockstone] point not so much to certain stars as away from the entrance of the outer ring where souls who were enslaved in form could find reassurance of the rites of embodiment. The distractions of the earthly manner caused erosions of behavior. Site was the reminder of the reason for embodiment: Thou shalt have the experience of flesh manifest in creation. At the far end of the Road site north lay a crocodile pen. These creatures symbolized greed, avarice. Atlanteans always referred to their origin in the stars for their understanding of the life forces which at low levels were negative in manifestation."

I was intrigued by Joan's confirmation of the Pleions' effect on human consciousness. She also added new details to Carol's dramatic picture of past earth changes affecting Bimini, and offered further clues to the magnetic problems to be encountered there.

In the distant past, Joan's readings suggested, the Pleiades rose much higher (closer to north), and then shifted gradually to the present easterly latitude. For a time they did rise at 45 degrees azimuth. The early astronomers at Bimini were interested not only in the rising and setting of the Pleiades but also their meridian transit or passage overhead. This was observed with a standing column similar to Cleopatra's Needle in Egypt. There was great concern over these astronomical observations because violent earth changes of the past had produced a very different night sky, and they were vigilant to recognize shifting patterns in the sky.

In 28,000 B.C., the Bahamas had undergone a 7-degree plate rotation when the upper layer of the earth's crust slipped on the lower. Could this be the explanation of the 7-degree separation between the two parts of the Bimini site? According to the readings, it *was,* as each part was constructed at a different time—one before, and one after the shift.

Both Carol's and Joan's psychic readings proposed a drama of earth changes and a history of heavy seismic activity that certainly runs counter to the known geology of the Bahamas.

Our present understanding of Bahamian geology has led to the impression that the archipelago is quite stable, geologically speaking. On the other hand, the Florida Straits and the Gulf Stream, whose depth between Florida and the Bahamas exceeds 3,000 feet, are seen by some authorities to be a result of a down-faulting process. At least one geologist proposes that the Gulf Stream between Bimini and Florida was a fault resulting from an ancient earthquake.

Geologist Robert S. Dietz, writing in *Sea Frontiers,* seeks to account for the Bahama plate's geological origins in terms of plate-tectonic theory. Dietz says that before Africa and the Americas parted, the Bahamas were once the center of a geological hotspot. He sees the Bahamas as a center of high lava flow, with evidence of this volcanism to be found below the present coral reef making up the several Bahama banks. As earlier noted, however, the basement rock is very deep, perhaps over 14,000 feet. Dietz also says that the hotspot that caused the lava flow under the present Bahamas was active until recent geologic times on the mid-Atlantic Ridge.

There were also unconfirmed reports from pilots flying between Miami and Nassau of a large fault in the sea bottom. Since the first edition, this fault has been confirmed by the United States Defense Mapping Agency. The three seabed fractures we found also hint at a somewhat more active geological history. However, the general geological evidence suggests nothing like the cataclysmic loss of a continent. If the Bahamas were the site of an ancient culture, as we contend, its demise was likely more gradual; it would have been inundated by the last stages of the Ice Age and, as the psychic readings suggest, may have also experienced a lateral shifting of the earth.

Once again, I found my respect for Velikovsky's theories of earth cataclysms being reinforced. If he is right, as I suspect, astronomical assumptions would be inaccurate for sites built prior to his date of 1450 B.C. (Venus' passage close to earth). Recalling Edgar Cayce's date of 10,700 B.C. for the final Atlantean destruction, we begin to see the folly of trying to make astronomical alignments for a site whose location may have shifted.

After much research during the winter, during a reading on April 17, 1976, I asked Joan this question: "Can the following list of the functions of astronomical orientations at sacred sites be expanded: (1) a calendar for agriculture; (2) a key to navigation seasons; (3) the conservation of the priestly power; and (4) a memorial of human origins?"

Her reading revealed the following: "Energies in horseshoe shape were collected and focused within certain boundaries. . . . These astronomical alignments contribute greatly to the energy and wholeness of the planet as a living entity with its own balance, own energies emanating from within and reinforced from without. Harmony, musical harmony from the planet could be detected. Then misalignments, cataclysms, the breaking and tearing. . . . Great monuments caught the resonances of power at certain points, thereby becoming focal points of terrific magnetic energy in the environment of which everything prospered. Within the temples men recharged as soul bodies generally, and specific healing took place. . . . The magnificence of those times was in their understanding and using harmonious influences to benefit the earth experience and the creation in general. Such magnanimous purpose is so foreign to today as to be practically inconceivable."

The allusion to the "resonances of power at certain points" obviously refers to the planetary grid system, wherein ancient temples were constructed on the power points of the grid.

In 1979, the year after the first edition of the present work, I published my global survey of megalithic sites that were clearly located on points of the Russian planetary grid system. This was *The Ancient Stones Speak.* In this work, I now see that I was only taking the first steps toward the insights of Blanche Merz, whose *Points of Cosmic Energy* (English translation 1987) develops a new discipline, geobiology.

In her book, she explains complex subsystems within the planetary grid system, including Ernst Hartmann's radiation grid, which covers the globe with north–south and east–west walls of measurable radioactivity forming rectangles 2 by 2.5 meters, except where modified locally by running water underground or other geological phenomena. Depending on an individual's physical location in relation to this grid, he or she will experience either positive or negative physical and psychological consequences.

Merz documented local modifications of the Hartmann grid with Geiger-Muller counters, magnetometers and sophisticated dousing instruments at such locations as Chartres and the Great Pyramid at Giza. Her work dramatically reinforces the long-held idea that ancient builders deliberately located their sacred structures globally on locations of positive energy.

This research clearly supports Joan's reading of 1976, in which she also predicted that magnetic anomalies on the order of 35 degrees would be found at Bimini today and that these anomalies, somehow related to solar winds, affected the weather and ocean currents. This reminded me of the unusual magnetic activity we found during the 1975 expedition. Consulting my field notes for that time, I began to see a connection between more extreme activity during the summer solstice. Later in 1976 we did establish that, whatever their explanation, magnetic anomalies do exist on the Paradise Point site. As yet, however, we have not observed any anomaly on the order of 35 degrees.

As I thought about the idea of an ancient influence from the Pleiades, one major problem loomed above all others: the nature of these beings. But as I began fitting together all the pieces—Plato's myth, Cayce's readings, Velikovsky's astrophysical theories of earth changes, as well as findings by contemporary archaeologists, geologists, physicians, parapsychologists, psychics, and my own field work at Bimini—the following conjectural scenario began to emerge.

In about 30,000 B.C. a multiplane emigration took place from the Pleiades to this planet. These emigrants did not require the hardware suggested by Erich van Däniken; indeed, they did not descend completely into physical form for perhaps thousands of years. Surely these beings' actual appearance on Earth would have convinced the resident population that gods were visiting. Closer to energy than matter, they probably manifested themselves somewhat in the manner of individuals being "beamed down" and they may possibly have been the Shining Ones of esoteric tradition.

(One problem having to do with research in this area is the difficulty of distinguishing between OOBE [out-of-body experiences] and clairvoyance. At present the American Psychical Research Society in New York is at work on the problem.)

The Pleions, closer in form to energy than matter, built temples designed to raise human consciousness, heal physical ills, and remind us of our origins—which were outside this planetary system. These temples were constructed with what John Michell calls a sacred geometry—their architecture employed certain sacred numbers related to the structure of the universe (which was well known to them), the solar system, and life on this planet. These temples were also oriented to favor stellar, solar, and lunar alignments. The subtle energies of these bodies resonated within the geometric shapes of the temples, and were thereby amplified, both for the raising of consciousness and healing. The temple's beneficial effects were enhanced by locating it in a significant relation to the earth's magnetic fields. The best known remaining temple that embodies these criteria is the Great Pyramid at Giza. Finally, the subtle effects of the temple were enhanced by sound—in other words, chants. For thousands of years the beneficial effects of these temples continued; then, through earth changes, most of the temples were destroyed. Small groups of initiates retained versions of this knowledge but gradually lost any practical influence upon humankind. The remains of one of these temples (perhaps, as Egerton Sykes says, the Temple of the Transparent Walls and the Golden Gates in the complex called Murias) was located at Bimini. Now tens of thousands of years later, its remains are almost hopelessly jumbled, but some of its original pattern still remains. By applying enough skill and energy, perhaps we will be able to unravel the pattern. It may be that, like Arthur Clarke's monolith buried on the moon in his film *2001,* these ancient stones are once again about to speak to humankind.

This hypothesis could shed light on several topics in prehistory. First, it would help to account for the widespread ancient interest in the Pleiades, which was far more intense than observational astronomy would warrant. It also offers an additional dimension to human evolution on this planet: we could have been assisted from time to time by extraterrestrial influence. Finally, it is consistent with esoteric traditions of nonmaterial beings, or souls, who came gradually into form. Indeed, in the months following the expedition, I found abundant evidence to show that the Pleiades exercised a widespread influence on ancient human consciousness.

According to Carol's readings, they had come in as energy, as light, only partially solidified into matter. In principle, this is consistent with accounts of human origins found in Plato, Plotinus, and Cayce. Today, the widespread interest in people as spiritual beings can be observed on the shelves of any bookstore, in subjects ranging from life after death to out-of-body experiences. Accumulating medical experiences, highlighted by Dr. Elizabeth Kübler-Ross in her work on death and dying, is providing evidence that consciousness exists as an independent energy system outside of the physical body. In 1973, while giving a paper at the American Psychiatric Association in Honolulu, I heard Dr. George Ritchie, a Charlottesville, Virginia psychiatrist, describe his own out-of-body experience during nine minutes of his recorded medical death. Ritchie is one of the few physicians I know with the courage to tell of his experiences while out of the body.

The idea of consciousness being able to function away from the body was recently investigated by Dr. Hal Puthoff and Russell Targ at the Stanford Research Institute. Their subject was Ingo Swann, who claims to be able to project his consciousness at will. Earlier, with the famous psychic and researcher Harold Sherman, Swann had projected his consciousness to the Jupiter environment *before* the arrival of the space probe directed to Jupiter. Significant parallels emerged later between Swann's and Sherman's accounts and the telemetry data returned from the space vehicle. During an out-of-body-experience at Stanford, Swann demonstrated his ability to affect the intensity of a magnetic field as it was being recorded by a well-shielded magnetometer. Further, Swann projected himself within the complicated magnetometer and gave a description of its complex internal mechanism—as if inside it.

12

LOOKING AT THE LEGENDS

I was aided in this research by a very helpful correspondence with a doctoral candidate in archaeology, Jon Douglas Singer, who had joined us on the 1975 expedition. It soon emerged that the Pleiades were the focus for an incredibly diverse series of facts in history, mythology, anthropology, and archaeology. What follows is only the tip of the iceberg.

The ancient Chinese, it seems, revered the Pleiades as early as the twenty-fourth century B.C., and we can even find reference to the Pleiades in the Bible, when the Lord challenged Job: "Canst thou bind the sweet influences of the Pleiades?"

The association of these stars with goodness was further highlighted in one of Joan's readings, which stated: "The Pleiades in mythology occupy an important place for a reason. An outthrust of universal love came from their midst which would be vaguely understood by earthmen."

Strangely enough, while brighter groups of seven stars are to be found in the sky, such as the stars of the constellation Ursa Major, ancient peoples throughout the world referred to the Pleiades as The Seven. For example, in Polynesia in the Pacific, the constellation is known as The Seven Little Eyes of Heaven. Furthermore, all over the world they have been regarded as seven, despite the fact that today only six are visible to the naked eye.

The Greek legend of the Pleiades, here recounted from *The Library* of Apollodorus (second century B.C.) is as follows: "Atlas and Pleione, daughter of Okeanos, had seven daughters called the Pleiades, born to them at Kyllene in Arkadia, to wit: Alcyone, Celoeno, Electra, Asterope, Taygeta, and Maia . . . and Poseidon had intercourse with two of them, first with Celoeno, by whom he had Lykos, whom Poseidon made to dwell on the Islands of the Blest, and second with Alcyone . . ."

A century later the legend was added to by the Sicilian Diodorus. He told of the Atlantioi, a people who warred with the Amazons. The Atlantioi's name came from their leader, Atlas, who also named the greatest mountain in his country after himself. Atlas was said to have perfected astrology and to have first given humankind the idea of the sphere. Diodorus also says of the Pleiades that they "lay with the most renowned heroes and gods and thus became the first ancestors of the larger part of the race of human beings. . . . These daughters were also distinguished for their chastity, and after their death attained to immortal honor among men, by whom they were both enthroned in the heavens and endowed with the appellation of Pleiades."

Ancient Greek temples were built to face toward the

Pleiades at their rising or setting, and in the classical world, the morning rising of the Pleiades was used to signal the summer's beginning and their morning setting the onset of winter. Thus these appearances also described the navigation season, which for the ancients was from May to November. Although the Pleiades are not used for celestial navigation today, curiously, the Arabic names of two important navigational stars in the area of the Pleiades are related to the cluster. One is Mirfak in the constellation of Perseus, meaning in Arabic "elbow of the Pleiades." The other is Aldebaran (the eye of the bull) in the constellation Taurus, which in Arabic means "follower" (of the Pleiades). Again I was struck by the recurring influence of Aldebaran and the bull, first in connection with the Bimini Road site, now as an important marker for sailors.

Writing in *Nature* (December 1881), R. G. Haliburton claimed that the worldwide traditions respecting the Pleiades go back into an unknown period of humankind's history when the constellation represented the "central sun" of religions, calendars, myths, traditions, and symbolism. Haliburton also claimed that the ancients "believed that Alcyone of the Pleiades was the center of the universe, and that Paradise, the primaeval home of our race and the abode of the Deity and of the spirits of the dead, was in the Pleiades . . ."

Quoting Muller, a contemporary writer on the religion of the Dorians, Haliburton says that "the famous eight-year cycle, which was in general use in Greece, was lunisidereal, and regulated by the Pleiades, and that the great feasts of Apollo at Delphi, Crete, and Thebes, were arranged by it. . . . Apollo, generally assumed to have been essentially a solar deity . . . was a god of the Pleiades, and hence the seventh day was sacred to him and others."

Early in his researches, Haliburton started with the

fact that in Peru both Peruvians and Christians observed the feast of the dead on the same day, November 2. In Peru, an annual ceremony of sadness called the Ayamarca occurred in November. It was a period of penitence and sorrow held just before the sun reached the highest point in the southern hemisphere sky. Haliburton felt that there must be some other astronomical reason for this, and suspected the rising of the Pleiades because of the sacred nature of the number seven (Seven Sisters) in both New and Old Worlds. I myself had noted the widespread use of the meridian passage of the Pleiades at Hallowe'en as the beginning of the calendar for a number of ancient peoples. However, the Pleiades had not risen heliacally for at least 12,000 years. In his subsequent search for ancient calendars, Haliburton located a very ancient Brahmin calendar in Tirvalore whose name for November was Kartica (the month of the Pleiades). Later he found that the Polynesian calendar was determined by the rising of the Pleiades at sunset, or by their being visible all night.

In *The White Goddess,* Robert Graves described the Pleiades as a calendrical device for the priests of various ancient sun cults. The classical writer Diodorus wrote of a people called the Hyperboreans, who settled on an island west of Celtic Gaul, which Graves says was Britain. The chief god of the Hyperboreans was Apollo, who had a temple in their main city. Apollo visited this island every nineteen years, dancing all night to harp music. (There are stone circles in Penzance, Cornwall, that have nineteen posts.) The time of his stay was from the vernal equinox until the rising of the Pleiades. According to Graves, Pliny stated that this year began in July.

Significant orientations of ancient structures to the Pleiades include the lines on the plains of Nazca in Peru, the Mississippi mounds, and early churches in the south of England, as well as a stone circle called Boscawen-Ün near Penzance. The old Druids also held the Pleiades in high regard.

Callanish, a megalithic site in Scotland, has also been found to be aligned with the Pleiades. The reason for such alignments, according to Anthony Roberts in *Atlantean Traditions in Ancient Britain* is that "the Pleiades were associated by the ancients with great cataclysms and inundations that were said to have shaken and drowned the world. They were also recorded as being the dwelling place of certain Giant Sky Gods who once visited the earth and conversed with early mankind."

Surveying the Great Pyramid at Giza, the Scottish astronomer Charles Piazzi Smyth looked for passages in the pyramid through which one could sight directly overhead the meridian passages of the stars. He realized that many ancient peoples began their new year at Hallowe'en when the Pleiades join the equinoctial point directly overhead. Smyth calculated that the date of construction of the Great Pyramid could be deduced from the time when the Pleiades and a circumpolar star, Alpha Draconis, were both on the meridian at midnight of the autumnal equinox. This date was 2170 B.C. The conventional date of construction is circa 2644 B.C. Although his work was at first ridiculed, he seems to have been on the right track. According to Dr. Marion Popenoe Hatch, University of California anthropologist, the meridian passage of the Pleiades was a phenomenon of importance to the Egyptian and other Mayan cultures. This is revealed in surviving Mayan writings as well as architectural orientations favoring the star cluster. The *Popul Vuh,* the creation myth of Quiché Maya, relates that 400 heavenly youths returned to the Pleiades after fighting with humans and suffering degradation on earth.

The megalithic site at Teotihuacán near Mexico has also been found to have alignments with the Pleiades. Teotihuacán at its height was an immense city 4 miles across. One of its major streets pointed west, toward the setting of the Pleiades (known to the Mayas as *tzab,* the rattle of the rattlesnake). Carved in the southeast face of a hill, as well as in the floor of a house on Teotihuacán's main street, are symbols resembling surveyor's marks. The line between them is parallel to the Sirius–Pleiades axis. A river was rechanneled to run along this line parallel to the street.

A possible reason why such astronomical events were important enough to ancient peoples to initiate a calendar can be seen in one of Joan's readings: "This path [the meridian passage of the Pleiades at Bimini] meant the regular continuation of the seasons. Many feared this passage—that of the Pleiades would cease because of great earth changes in the past." The seasonal behavior of the Pleiades was, it seems, a sign of order in the cosmos, a sign that the awesome earth changes of tradition were not about to be repeated. This idea was dramatically reinforced for me when I came across an ancient Aztec ceremony connected with the Pleiades.

The Aztecs used two calendars simultaneously. The first calendar, which determined the seasons, was of 365 days, based on the sun's movement around the earth. The second calendar, used only for religious and ritual purposes, was based on a 260-day cycle. Each day of the year was considered in terms of two independent calendar cycles. Since the solar and ritual years were of different lengths, the two calendars coincided only once in fifty-two years. These fifty-two-year periods had great significance; at the end of one of them the universe was due to be destroyed by earthquakes. Nobody could foretell which epoch would be the last.

The last day of each fifty-two-year period was a time

of crisis and fear. Would the sun ever rise again? On those evenings, the priests would go up to a mountain top near Tenochtitlan to watch the stars. When the Pleiades had reached their zenith, a prisoner would quickly be stretched out over the altar, his chest would be opened, and the astronomer-priest would kindle a fire in the victim's body cavity. Messengers would light their torches from this flame, and run to carry the New Fire to every temple and every home in the Valley of Mexico. Life would continue for at least another fifty-two years.

In *Myths of the New World,* Daniel G. Brinton explains the sense of awe many early peoples of the New World felt for the Pleiades: "Aka-kanet . . . in the mythology of the Araucanians, is the benign power appealed to by their priests, who is enthroned in the Pleiades, who sends fruits and flowers to the earth, and is addressed as grandfather." A California tribe reverenced the Pleiades "to such a degree that to look at them carelessly was calamitous." And in Peru "they were worshipped as first of the starry host."

The influence of the Pleiades is to be seen even in the remote jungles of Brazil. Claude Lévi-Strauss, in his important work on mythology, *The Raw and the Cooked* (English translation, 1969), tells of the Sherente, who count the months by lunar phases and start their year in June with the appearance of the Pleiades, at the time when the sun is leaving Taurus. June begins a four-month dry season for the Sherente who count thirteen months.

The Hottentots of southern Africa celebrate the cult of their supreme god, Tsui-Goab (Wounded Knee) when the Pleiades appear. Tsui-Goab commands storms, sends rain for the crops and speaks with the voice of thunder. In Samoa in the South Pacific there is a sacred bird called Manu-lii, which Haliburton translates as the bird of the Pleiades. To the northeast of this island group lies Danger Island where, in 1857, the natives were discovered greeting the Pleiades with religious joy and feasting.

On Easter Island, usually considered the only megalithic culture of the Pacific Ocean, are found giant statues weighing up to 90 tons. Following the original megalithic culture, a Polynesian culture came with the Birdman fertility rite, which times the beginning of the growing season and coincides both with the rise of the Pleiades and the arrival of the sooty tern from the Marquesas and other islands to the west.

Why, we may finally ask, did ancient people pick a relatively obscure star cluster to signal the most important events of life? In my opinion these myths, legends, calendars, and structures oriented to the Pleiades' rising indicate profound and widespread religious emotion, arising from the racial memory of a spiritual contact with the star system centuries before.

Early in my own work on the topic, a former graduate student, David Cammack, brought me a translation of the *Kumulipo,* a Hawaiian creation prayer written down about 1700 and chanted to Captain Cook during his visit to the islands.

CHANT ONE

At the time when the earth became hot
At the time when the heavens turned about
At the time when the sun was darkened
To cause the moon to shine
The time of the rise of the Pleiades
The slime, this was the source of the earth.

translated by M. Beckwith

I will always remember the impact of those lines as I suddenly became aware of a possible connection between the 28,000 B.C. date of the first influence of the Pleions at Bimini from Carol's readings; Joan's reading, which described an earth change in 28,000 B.C.; and the verified fact that the earth's magnetic field did reverse in 28,000 B.C. The line "at the time when the heavens turned about" carried a theme of destruction whose implications were more than I wanted to consider.

In *The Ancient Stones Speak* (1979), which surveyed the megalithic sites of the globe, I was led to study these sites because of the worldwide distribution of legends connecting the Pleiades with ancient natural disasters on earth, of which I've just highlighted but a few: the Hawaiian creation legend, the *Kumulipo,* just described, in which creation is born out of cataclysmic destruction from global fires and flood; then Beltane, the Celtic May Day, signalled by the rise of the Pleiades at dawn; and these compared to the Aztecs' fifty-two-year cycle timed by the midnight meridian transit of the Pleiades, as well as to the calendar of the Sherentes of Brazil, who mark their seasons by observing the Pleiades.

Even more compelling than this for me, however, is the worldwide distribution of ancient sighting arrangements favoring the Pleiades, which have been uncovered by archaeoastronomy at various megalithic sites. Some of these sighting arrangements linked with the Pleiades include Teotihuacán (setting), the Acropolis (rise), Callanish (rise), and Namoratungu (rise). As earlier mentioned, these megalithic sites also were located on intersections of the planetary grid system.

But again, there is more. Although I had long been interested in native American cultures, it was only in 1988 that a friend introduced me to a book based on the Cherokee teachings by Dhyani Ywahoo. I was astonished to read of a tradition that went back 100,000 years and described the Tsalagi (Cherokee) as descendants of star people from the Pleiades, or the Seven Dancers, who first came to Atlantis, or Elohi Mona, to help humankind evolve mentally. After the destruction of Atlantic, in the Tsalagi tradition, the survivors came to Turtle Island, the present United States.

Aside from the astonishing reinforcement of the Pleiades' influence in prehistory, I was also surprised to read that the Tsalagi tradition also held essentially the same view of the reasons for the destruction of Atlantis as did Plato and Edgar Cayce: greed and lust, or, essentially, the manipulation of others by power-hungry leaders for nought but their own personal gain.

PART IV

ON BOARD AGAIN: POSEIDIA '76–'79

13

POSEIDIA '76

The fall of '75 passed very slowly. I felt the painfully acquired momentum slip away as I found myself engulfed in paper work the expedition had created—a radical change of pace from the previous summer's activities. During this time I chanced upon the words of the British explorer Col. P. H. Fawcett, written while waiting to return to the jungles of Brazil in 1923: "At the time of writing these words, I am awaiting with what patience I can muster the culmination of plans for the next expedition." My sympathies went out to him.

From California John Steele wrote me of his efforts to secure opinions on our "building block" from various pre-Columbian archaeologists at UCLA and Berkeley. An expert on old Mexican artifacts showed John a photo of a similar-looking stone that was located at the famous Olmec site at La Venta.

A physicist friend at the Brookhaven National Laboratory on Long Island was unable to date the block by standard thermoluminescence techniques because it had not been fired. However, Dr. Edward V. Sayre, the head chemist at Brookhaven, confirmed that its composition was a sandstone-limestone mixture and suggested that it might have been created by an ancient technique of mass production! I assume his opinion was based upon the fact that this block's flat sides are not parallel and that the mineral components might possibly have been cast in a mold.

At the beginning of 1976, I was trying to unravel all the clues found at Bimini when an important break came. Frank Auman, a North Carolina businessman, introduced me to Karen G., the only psychic to be hired on a full-time basis by Dr. J. B. Rhine, the foremost pioneer in American parapsychology. Many psychics are effective healers and counsellors; some are used in police work. Karen has served in all of these capacities, but was presently involved in an experiment that to me seemed even more impressive than the work of Uri Geller. For three years she had been bending an ultraviolet beam an amazing 45 degrees by mental concentration. The experiment is respectably situated in the Electrical Engineering Department at Duke University.

Karen expressed an interest in the Bimini project, so I showed her the tongue-and-groove building block before taking it to the museum in Nassau. Holding it in her hands, she began to perceive an ornate temple located, she said, about three miles northwest of where I had found it. Later Karen drew the temple in rough outline. So far, its form has been tentatively related to an Egyptian temple.

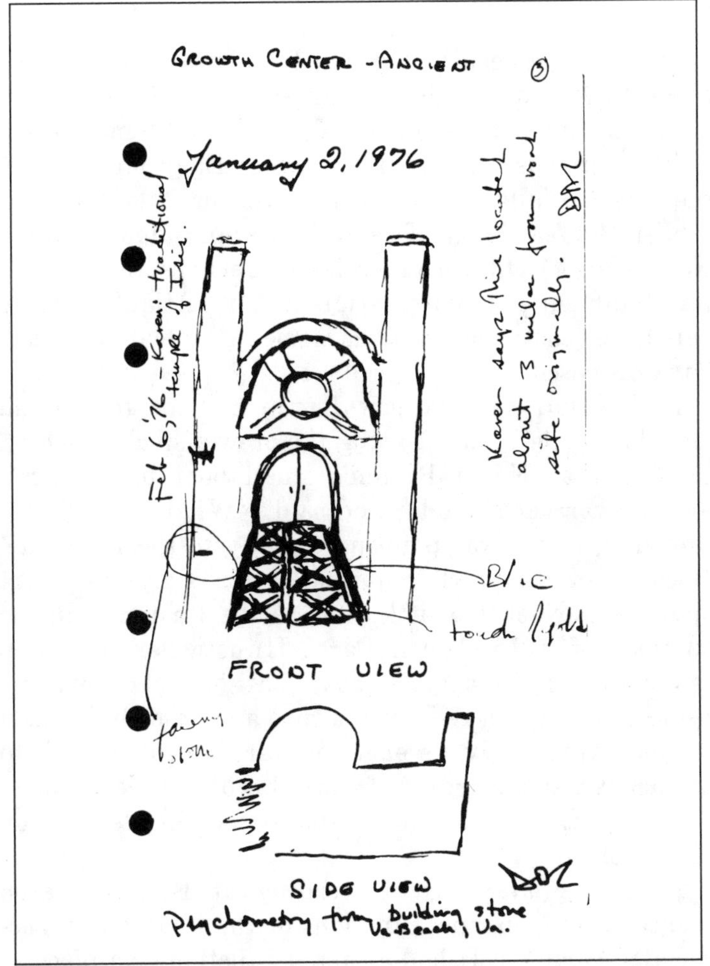

Psychic Karen G.'s impression of the tongue-and-groove building block discovered during Poseidia '75.

This psychic clue reminded me of Egerton Sykes' theory that after the fall of Atlantis, first the Egyptians and then the Phoenicians visited Bimini. Both Karen and Sykes thus shared the idea that Bimini had once known an Egyptian influence. Could this, I wondered, have been the result of an early transatlantic voyage from Egypt, a possibility suggested by Thor Heyerdahl's *Ra* adventure? (He had, it will be remembered, built a papyrus boat from an ancient Egyptian pattern, then successfully navigated it across the Atlantic to establish the feasibility of early Egyptian influence in the New World.) On the other hand, perhaps the "Egyptian" influence actually originated in Atlantis before spreading east and west, as some Atlantean theorists have proposed.

In February Frank flew Karen and me to Bimini. One day while I was showing the crew a small beach off the point at North Bimini's small boat pass, Karen looked over her shoulder and said, "What's that?" Astonishingly, she was pointing directly to the East Site. Because of the work required at the Paradise Point site, we had devoted little time to the East Site in the summer of 1975. When Carol Huffstickler had been taken over it in a small boat, she had clairvoyantly heard singing of a religious nature and described it as a temple. Apparently its energies were potent enough to attract Karen's attention from a beautiful scene immediately in front of her to an area barely visible to us on the beach.

During a later psychic reading at Bimini, Karen identified the multiple functions of the Road and its associated complex. It had a sacred function as a place of prayer and spiritual energy, but was also a trade center, and a "gateway." Although this last function was not clarified in the reading, I had the impression that perhaps, due to the heightened spiritual energy of the site, out-of-body experiences may have been facilitated there. In response to a question about the influence of the Pleiades at Bimini, another somewhat cryptic reading said that "transportation" to the site (from the Pleiades?) was "via mind and light." As in Carol's readings, hardware such as UFOs did not seem very important; but again there was a connection with the Pleiades.

One of Karen's more intriguing comments about the Road site was her statement that its overall configuration was that of a trident. If this could be established, it would certainly be a strong link to Neptune, or Poseidon as the Greeks called him.

I gave Karen the marble chip from the stylized head which, at that time, was still resting on the seafloor. Her impressions gained from holding it included the notion that it was a "tribute marker" from another culture, and she saw it sitting atop a pile of stones arranged like a totem pole. (A strange coincidence occurred some months later when I was a consultant to the Cousteau Society at Bimini. Without telling them about Karen's reading, I showed them the head underwater. Before filming it, they stacked the head atop the three other marble blocks nearby!)

Several days after our arrival, I took Karen on her first visit to the Paradise Point site. Because of the rough seas that day, I myself was uncertain of the location. When I anchored to get my bearings, Karen told me that our present location had good vibrations. Once in the water, I found that we were anchored within 20 feet of the chockstone, that crucial point in the Road survey that suggested a sacred geometry. Now that it had brought forth an independent response from Karen, I began to feel better about the project.

Our activities at Bimini had attracted the interest of a Florida organization called the International Explorer Society, who began to consider giving me the award which they had earlier presented to the Peruvian archaeologist, Dr. Maria Reiche, in 1975. Dr. Reiche had spent twenty years working on deciphering the strange markings on the plains of Nazca in Peru, and felt that some of the lines had astronomical significance. Nazca's rectangle (the largest geometric shape) pointed to the rising of the Pleiades between A.D. 400 and 500. Gerald Hawkins later further refined this to A.D. 610. I was gratified to discover yet another indication of the Pleiades' influence.

About a month later I learned that the society had scheduled a press conference on Bimini Island during April, at which time they planned to name me Explorer of the Year. At the press conference I explained my reasons for believing that Bimini held a megalithic site with a possible sacred geometry. Wire service stories about the conference led Helio Costa, New York bureau chief of Globo Televisión—the Brazilian national network, with 15 million viewers—to interview me at Virginia Beach on the Bimini project and its background.

During the summer in the midst of work on this book, I received a telegram that began: "The Cousteau Society filming on Bimini needs expert on submerged formation to join them immediately." Jacques' son Philippe was on location at Bimini filming for their "Lost Civilizations" special, and needed a consultant for the Bimini site. I agreed to leave the following day. Philippe proved to be an extremely competent aircraft commander and expedition leader. The work was accomplished quietly and professionally from the flying *Calypso,* a converted Navy PBY seaplane.

In the air some miles east of Bimini, Philippe called me on the intercom to point out an astonishing sight. He had prepared me, before takeoff, but what I now saw was impressive beyond words. Below us ran straight lines of circular white spots on the seabed; the previous day, diving, the *Calypso* team had found the spots to be areas clear of sea grass. Suspecting fresh water, they had taken water samples but had had no way of running salinity tests aboard. These amazing spots ran in a perfectly straight line for miles, not to be confused with the mysterious White Water of the Bahamas, said to be caused by bottom-feeding fish. In one place, three such lines intersected each other. I had never heard reports of this phenomenon in the Bahamas and for some reason had never flown the track we were now taking to Andros Island. Later the following day, on a flight to the southern edge of the bank, we found another line of these spots. Still later, an oil company executive provided a possible explanation: the explosions of shallow-water seismograph work had disturbed the bottom. Once again the mysterious waters had presented us with an enigma, and we added it to the growing number of unsolved questions about the area.

One of the sites that Philippe wanted to check out was the so-called Andros Island "temple." Since the site was in shallow water some distance offshore, the PBY was most appropriate from the point of view of logistics. When Philippe and I dove on the site in 4–5 feet of water, we found that it was as an island informant had earlier claimed—a holding pen for sponges. There was absolutely no sign of sophisticated masonry, only roughly shaped coral pieces stacked up about 18 inches above the seafloor. Yet another story, however plausible, collapsed. This survey by the Cousteau Society was later a part of "Calypso's Search for Atlantis," aired on KCET in Los Angeles.

The following month I organized the fifth expedition to Bimini, Poseidia '76, which was supported by a chartered 41-foot diesel auxiliary sloop *Sea Fiddle,* out of Miami. My staff included a retired Navy captain, John "Josh" O. Sherman, Jr.; Douglas G. Richards, a Ph.D. candidate in zoology; and Brady Chapin, a geology student. Josh, an aviator with an unusually wide range of experiences, often acted as an adviser for my explorations; this time he handled Poseidia '76's survey from the beach as well as keeping track of our logistics. Doug

The author at the helm of the ketch Sea Fiddle *during Poseidia '76.*

and Brady also contributed from their respective scientific disciplines.

It was an expedition with limited but nonetheless important objectives. First I wanted to pin down the strange magnetic activities that I suspected on the Paradise Point site. Second, I wanted a more extensive series of geological samples from blocks facing each other, in order to learn more about their jointing patterns. Then there was a need to refine our aerial survey. But above all, I hoped to raise the marble head.

Our first objective was easily met. The general orientation of the site, as observed in August of 1975, was 45 degrees magnetic (or 43 degrees true, due to the 2 degrees westerly of local variation). During the summer, these figures had deviated as much as *15* degrees from what seemed to be the most valid figure in August. The greatest variation had apparently occurred around the time of the summer solstice in late June 1975. We suspected magnetic anomalies, of course, but lacked the equipment to verify our suspicions. This time we set up a baseline on the beach with a transit, observed the sun for the true bearing of our baseline, and then, by triangulation, established the positions of key survey points buoyed on the site a half mile offshore.

This expedition established two facts: (1) that at this time (August 1976), the local variation (the difference between true north and magnetic north) on the beach was as charted, or 2 degrees westerly; and (2) that an important section of the main lead was oriented 49 degrees true.

This meant that in 1975, at the same time of year, we had apparently experienced a 6-degree magnetic anomaly. What our findings meant so far was that, in addition to the local (charted) 2-degree westerly variation, we had observed an additional 6-degree variation in August of 1975 and an additional 3-degree variation in August of 1976. These observations added to my growing awareness of unexplained magnetic problems in the Bahamas. I suspect these are caused by the position of astronomical bodies, but any attempt to explain such

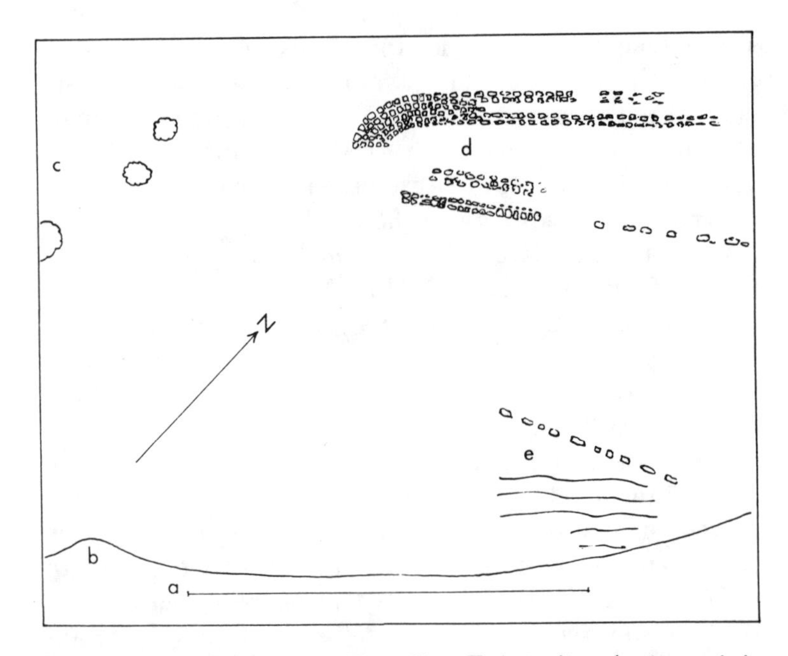

Key survey locations of Paradise Point site. (note: not to scale)

a. 1200-foot baseline for 1976 survey
b. Paradise Point
c. Coral Crossing Rocks
d. The Road site
e. Stone alignment recently seen from the air but not yet investigated, which does not parallel the ancient beach line (wavy lines below). Highly convincing evidence for human construction of the site.

anomalies would lead us off into the limbo of such topics as the Bermuda Triangle.

Additional samples were taken from the stone blocks and analyzed by Brady Chapin for course credit at his university. The results were inconclusive.

Our final two objectives were aborted by tropical storm Dottie. The bottom visibility from aloft was poor, so that aerial photography was useless. What was even more dismaying was that after the storm we were unable to relocate the marble head. I hoped that it was still somewhere in the waters, hidden by sediments Dottie had stirred up.

While returning the *Sea Fiddle* to Miami, for the first time in my years of blue-water sailing I had occasion to call the Coast Guard for assistance. We had been plagued by fuel problems all week. Then, when we needed the diesel just off Miami, it stopped. This time I was unable to start it. The charter was up in three hours, so I called the Coast Guard; they had us in tow within an hour. Their timeliness and courtesy on this occasion more than made up for the hazing we'd gotten the previous year in the Gulf Stream.

Four months later, in December of 1976, a news story caught my eye: "Soviet Links Moon, Bermuda Triangle." A. J. Yelkin, a Soviet engineer whose work had just been reported in *Izvestia,* had plotted moon positions against times of aircraft disappearances and found the greatest number of disappearances at the time of the new moon, full moon, or when the moon was closest to the earth. His conjecture was that "lunar-solar tides may cause magnetic disturbances beneath the ocean, disrupting aircraft compasses and throwing an aircraft onto a fatal course." This connection of lunar-solar positions with magnetic anomalies in the area was very exciting! Our own findings in August had earlier suggested that magnetic anomalies at Bimini (in the Triangle) were perhaps greatest at the summer solstice in June, then declined toward August. We seemed to be on the right track.

14

POSEIDIA '77

Once again, a boat was towing me through the crystal-clear Bahamian waters in search of the marble head Gary Varney had found in 1975. For two years it had lain on the seabed off Paradise Point, while I made repeated efforts to relocate it for the museum of the Bahamas Antiquities Institute. I had seen it once again in the interval while diving as a consultant to the Cousteau Society. Philippe Cousteau was unwilling to become involved in raising it, however. He and his divers photographed it, then went on to other subjects.

In August of 1976, a friend in south Florida told me that a Miami smuggler had become interested in the artifact, thought he knew its location and planned to pick it up, then sell it! During the next expedition in October, passage of a strong early cold front destroyed underwater visibility, and for seven months I had to live with the possibility that someone else might get there before me.

During the previous winter I had received Brady Chapin's lab analysis of thin sections from various hand samples. Reading his report, it was difficult to say what conclusions we could draw about the site as a natural formation of beach rock *versus* one worked by man. In one location near the chockstone, hand samples from adjacent blocks showed different cementing patterns just as before in Parks' work (and also noted in one of Gifford's reports). The rest of Chapin's samples did not show this difference. Chapin therefore took the conservative position that the blocks seemed to be a natural formation. However, his limited number of samples did include one whose structure, as determined by microscopic analysis, was markedly different from the others, not only in its cementing but also in its constituent elements.

I was still not convinced that the entire site had been formed in place. As I reviewed all my evidence, I was left with two possible explanations: (1) the blocks had been cut elsewhere and moved to Bimini from an essentially homogeneous limestone bed (originally formed as beach rock); or (2) just as Bahamians cut coral blocks on land today, the blocks were cut into patterns *in situ,* with the addition of some foreign material such as the chockstone. Later, as the seas rose, the native blocks were hardened by a sea water exchange of calcium carbonate that recrystallized and consolidated the porous limestone.

With the advice of various marine geologists up and down the east coast, I found that the only procedure now open to me was to core adjacent blocks and compare the bedding planes of the two adjacent cores. Poseidia '77 had taken shape in the third week of June in southern Florida. With the help of Dr. Ray McAllister, professor of ocean engineering at Florida Atlantic University, we were able to use Miami-Dade Junior College's research vessel *Martech,* and to rent a hydraulic corer to take core samples from the blocks of limestone on the Road site.

Now, as I watched the bottom pass under my faceplate, I wondered if I had already been beaten to the marble head. Were we merely wasting time and fuel? Suddenly I thought I saw it, marine growth all but obscuring its square shape. I shouted to Kevin McAllister who was running the boat, then dropped off the tow line to dive to the bottom 20 feet below. I wanted to be sure. It *was* the head! I experienced a feeling of great relief; the head and three other rectangular pieces of marble were still there. Kevin anchored the boat. Peter Baas and I got into our scuba gear and dove down with two lifting bags, each capable of supporting 250 pounds.

I tried to temper my excitement as we lashed quarter-inch nylon line around the head, attached the two lifting bags, then began to inflate them from our regulators. We had filled both nearly two-thirds full of air without results when the critical balance point between the force of gravity and the buoyancy of the inflated bags was passed. The head lifted off and accelerated toward the surface. We left the other pieces of marble to mark the location.

As we motored back to the *Martech,* I stayed in the water with the head on another tow line. I had waited long enough and didn't plan to lose it again. Once at the stern of the *Martech,* the highly stylized head was easily winched aboard with the A-frame. On board, our prize elicited some skeptical comments from other members of the expedition, and I overheard someone say: "He claims it's marble," unaware that I had taken a small chip from it in 1975. Relieved at having salvaged the relic, I let them have their fun; I was beginning to feel much better about the expedition.

It would be three years before the marble head was finally picked up from the Bimini jail and delivered to Dame Doris Johnson in Nassau by Mark Hosey on March 22, 1980. It was then included in the holdings of the Bahamian Antiquities Institute.

And it was not until 1989 that I encountered any rational suggestion of a possible cultural relationship between the marble head and any other known culture. When doing a radio talk show in Leucadia, California, I heard from a caller, Ken Johnson, that the work of archaeologist Marija Gimbutas in the Balkans had discovered stylized heads that could possibly be related to the Bimini find. After the show Ken wrote me as follows:

> The sculptured head found off Bimini is said to bear no resemblance to Meso-American sculpture of any period. This is true enough. Some Meso-American sculptural traditions feature naturalistic portrait heads; the Bimini head is not naturalistic. Most pre-Columbian sculptured heads, however, are elaborate affairs which seem to be based on ritual masks; the Bimini head is far from elaborate. In fact, the features are so lightly modeled as to give the whole piece an abstract quality. At the same time, there is a monumental bulk to this piece which is characteristic of heroic statuary in many other early cultures.
>
> Can this combination of stylized, semiabstract facial features with heroic monumentality be found together in any other culture? I believe that it can. The answer, however, is not to be found in any of the cultures usually favored by Atlantean researchers, but in a totally unexpected quarter.
>
> During the past few decades, archaeologists have become aware that a very highly developed Neolithic culture developed in what is now southeastern Europe, the Balkan or Carpathian region. This culture, sometimes called Old Europe, dates back to 6500 B.C.—perhaps earlier. Several examples of sculptured heads from Old Europe—particu-

larly from the Yugoslavian sites of Lepenski, Vir, and Vinca—can be found in *The Goddesses and Gods of Old Europe, 7000–3500 B.C.,* by Marija Gimbutas (1982). These Balkan figures show the same monumental quality, as well as the same lightly modeled, almost abstract facial features.

Interestingly enough, archaeologist Marija Gimbutas, one of the principal excavators of Balkan sites, has linked the religious sculpture of that region—in terms of both style and content—with the civilization of Minoan Crete. And Crete, of course, is the culture that conservative archaeologists offer as the "original source" of Plato's Atlantis tale.

As if this were not provocative enough, the possibility exists that the oldest writing in the world comes from Old Europe. Since the Tartaria tablets were first discovered, there has been no end of controversy as to whether these inscribed marks on clay tablets do, in fact, comprise a script. The tablets *may* date back as far as 5000 B.C. If they represent an actual script, then archaeologists would be forced to admit that writing developed in the Balkans almost 2,000 years before it arose in Sumer.

Meanwhile, as Poseidia '77 continued, I was dismayed to learn from Ray that the core boring was not going well. He reported that even with a lifting bag attached to the hydraulic-drill motor, the 5-foot core barrel and bit could not be stabilized in a vertical position. Soon it became evident that we would have to engineer and fabricate a tripod for the drill, and the following day Ray supervised the building of one. In the week that followed, we took twelve cores from pairs of adjacent blocks.

For the first few days, the coring proceeded smoothly. Then one morning Ray and one of his students, Susan Kaplan, were operating the hydraulic drill about 18 feet beneath the *Martech.* Suddenly the 1-inch water hose that ran overboard to lubricate the bit was pulled over

Setting up the tripod for coring.

the side, then went slack. A few moments later Ray and Susan surfaced at the stern with tense faces.

We had been using a 5/8-inch nylon line and an anchor to fight the drill motor's tendency to twist out of the operator's hands whenever the bit caught in the rock. This time, the bit had really jammed. Ray said that the torque of the hydraulic motor, its control jammed by the breaking line, had at once begun to twist the sturdy hydraulic hoses around Susan, who was operating the drill. The sound of the line parting, like a rifle shot under water, had alerted him to look overhead. Moving quickly, he cleared the control and stopped the motor.

I realized that his quick response had probably saved Susan from very serious injury, perhaps even death. This was the most dramatic demonstration of Ray's underwater abilities. Each day he solved lesser problems in his stride, gave everyone an apprenticeship with the core drill, and in general was the source of overall goodwill aboard the *Martech.*

Back in Florida, preparing the second phase of Poseidia '77, I needed laboratory work on the cores we had taken and was looking for a way to fund it. Ray once again came to our rescue. He talked to Dr. Roy Lemon, the head of the geology department at Florida Atlantic University. I was delighted to find serendipity operating again. Beach rock was one of his own research subjects; he therefore agreed to fund the laboratory preparation of the cores and handle the analysis himself.

After about ten days ashore in Florida, the second phase of the expedition began to take shape in Miami, where I had chartered a 41-foot ketch for the next three weeks. Some of the 1975 staff returned, including John Steele and Gary Varney. My wife, Joan, had flown down from Virginia, and Joseph Libbey, diving consultant to the Smithsonian Institution, had arrived to give us the benefit of his extensive diving experience. Larry Arnold, a cartographer interested in megalithic sites, and Ronald Greening, our cook from Toronto, Canada, were also aboard. Others came and went in the course of the expedition.

We were motoring the 41-foot chartered diesel auxiliary ketch *Shepahoy* toward the Moselle Bank northwest of North Bimini, a bit west of North Rock Light. Overhead floated a few low, soft, fleecy clouds of the Bahamian summer; westward, over the Gulf Stream, the usual afternoon thunderstorms were building up. While we steered towards the Moselle Bank to investigate the supposed remains of another shipwreck, our guide, "Bonefish Sam" Ellis, entertained us with stories of the old days in the Bahamas. It had been hard to wrest a living from these islands and their sparse vegetation; moreover, the indolent climate did not inspire anyone to heavy labor. In these circumstances, it isn't surprising that during the latter half of the seventeenth century, the British colony's largest industry was wrecking.

The usual histories play down everything but the industry's salvage and lifesaving aspects, but even the bare facts are pretty suggestive. This industry, dating back to about 1648, provided a major basis for shipbuilding and gave the participants an official 40–60 percent of the spoils from wrecked ships. Suffice it to say that Bimini was settled (again) in 1848 by five wrecker families totalling forty souls. In 1856, according to Paul Albury's *The Story of the Bahamas,* about half of the able-bodied male members of the Bahamas (or about 1,300 in all) were licensed wreckers, and the number of licensed wrecker vessels, known for their speed, was 302! I found this astounding.

Sam filled in some of the more lurid details of this incredible business, including the practice of deliberately sinking passing vessels. At night a group of islanders would come alongside a passing ship that seemed to be making for the island, hail the ship, and offer to pilot it in. If the captain was agreeable, they boarded, whereupon some of them would disappear below decks, carrying with them large augur bits such as those used by shipwrights. As Sam put it, "Two an' a half turns, mon—right through the hull!" The sinkings were planned so that the vessel would go down near the island for easy salvage. Sometimes, Sam said, the captain was even in on the deal for salvage and insurance. How often this sort of thing actually happened and how much of it is folklore would be hard to say, but the community prospered enough that by 1881, according to Albury, the population of Bimini had risen to 663.

Upon investigating the granite blocks on the bank, we found what seemed to be modern quarry markings and decided to turn our attention to more promising areas. Then, on the return trip to Bimini, Sam told us that the Road site had once been constructed of blocks nine to ten stones high during his youth in the 1920s. In fact, he said that the site was charted in those days as shoal water—a hazard to navigation. When I asked what had become of the megalithic limestone blocks, Sam said that Arthur Sherman, a local businessman employed by a marine contractor, had removed them. Sherman later told me that he lifted granite only from the Moselle Bank area. Regardless of who removed the stones, Sam's story is supported by one of Joan's readings, which disclosed that "this entire site was one of immense proportions for it stood thirty feet high. Sea water washing over has eroded and destroyed silica strata, which razed the entire structure."

Rapid calculations confirmed that nine to ten of the medium-size stones would reach a height of about 30 feet. I added this information to my notebook, now growing bulky with corresponding data. After this expedition, Kiiri Tamm and I were to dive on the Miami Harbor breakwater. We surveyed the north breakwater from seaward to the beach. We found only quarried granite blocks, no megalithic limestone blocks.

The most intriguing objective on my agenda for this expedition, however, was a mysterious column Dr. William Bell of North Carolina had discovered in 1957. Bell was swimming off South Bimini when he came across this artifact which has baffled experts in subsequent investigations.

My own pursuit of this column began in 1973. I wanted to get a firsthand account from Bell himself, and, with the help of Frank Auman, arranged a long-distance conference call. During our conversation, Bell's excitement again revived as he recalled how he and other members of his party were diving in 40 feet of water when they noticed the column. It was about 4 inches in diameter at the top and 8–10 inches at the bottom where it penetrated the mud of the seabed. When Dr. Bell attempted to scrape off some of the marine growth covering it, he found the surface beneath to be covered with an unidentified gray substance. Just beneath the bottom mud he found a gearlike shape on the column about 2 feet in diameter.

Fascinated, he dug further and discovered another cog or gear shape about 3 feet deeper. Lying near the column were many granite slabs—perhaps twenty of them on the surface, with many more in the mud. Bell estimated their dimensions as 12–18 inches by 2–3 feet by 8–15 feet. Wondering what all this might mean, he decided to take photographs of the column and see what further clues he might get from studying them. He was shocked to discover that the developed photographs revealed peculiar radiation patterns—published for the first time on these pages. He had seen nothing during his underwater observations. I consulted with radiation physicists at Brookhaven and the NOAA (National Oceanic and Atmospheric Agency) who suggested that, through ionization, ultraviolet radiation might raise the energy of the water and become photographically visible without special filters and film.

Strangely enough, for several months after his contact with the column, he experienced a number of nosebleeds—completely out of the ordinary for him—that have never been explained.

Shortly after my talk with Bell, I was on Bimini for the February '76 expedition with Karen G. My hope was that Karen might be able to pick up a clue to the column's location. A strong cold front was passing through at the time, causing high winds and very lively seas. In a rented Boston whaler, Karen and I headed for South Bimini in the face of unmanageable seas. Finally clear, I let Karen direct our course, asking her to tell me of any interesting energy pattern she might sense. In a short while, she said: "This feels very good where we are."

I stopped the engine to take cross bearings on landmarks ashore. As I did, I was astonished to see that we were probably only about 100 yards south of one of the landmarks given me by Dr. Bell the week before.

On the beach a bit later, Karen saw clairvoyantly that energy of a white color was radiating skyward from the column's location. She also saw that there was more than one energy source, and that all of them focused into a ring elevated about 50 degrees above the sea! Then, to my further surprise, she saw a similar phenomenon south of the first and slightly seaward. Its color, clairvoyantly viewed, was pink. Perhaps Bell had chanced upon only one of many such columns.

The questions raised by Bell's discovery are awesome. Where did the column come from? If it is indeed radiating energy, what is the nature of this energy? And what is its source? Could this be related to the crystal of the Cayce readings that powered transportation and provided healing on Atlantis? I can say only that Dr. Bell's photographs, his description of his experience, and then Karen's response—apparently to the same phenomenon—all began to seem to me like science fiction come true.

Later I decided to ask Joan for information on the column and its location. Her response revealed the following: "This column is standing precisely where it stood years ago. Its purpose then was twofold: as an astronomical alignment point, yet contained within a building in which was housed a tremendous crystal. They worked together in certain ways—the column pointing out the sequences in the heavens about which the sun's energy could be magnified. The crystal reflected the energy of the sun, but it could be tuned more accurately at certain times of the day and year. The standing column dictated the seasons and though it was thought that man's soul emanated from above, the longing to be flesh was accentuated at certain times of the year. Feast days celebrated the times of turning out, times of less friction, times of harmony."

Joan's clairvoyant account was consistent with Cayce's crystal and its associated exotic technology, except that the column appeared to be an added refinement on Cayce's crystal, used for the reception and re-radiation of cosmic energy.

Now, during the 1977 expedition, I wanted to search for the column with the help of a side-scan sonar and a sub-bottom profiler. The latter, which can penetrate and locate objects within sea-bottom sediments, was invented by Dr. Harold Edgerton of MIT. Although he expressed interest in joining us on the expedition, at the last minute Dr. Edgerton was unable to come when the U.S.

First photographs ever published of the mysterious lighted column. Note the varying widths of radiation from the sides. (Courtesy Dr. William Bell.)

government asked him to attend a conference in the USSR.

By the time I learned of his cancellation, I had already committed myself to the expedition, so I immediately contacted Martin Klein, who had trained with Dr. Edgerton and had recently been off in Scotland looking for the Loch Ness monster with his electronics equipment (see *National Geographic,* June 1977). Klein, too, expressed a desire to join us, but his customers in the oil industry required his presence elsewhere.

My last hope was a Ft. Lauderdale firm whose owner had also been trained by Dr. Edgerton. Our expedition funds at this point were barely adequate for a twenty-four-hour hire of the equipment; since we were operating between Florida and Bimini, our actual on-site use would probably be only about six hours. We were still willing to proceed, under less than optimum conditions, but were unable to engage a work boat with a generator large enough to handle the profiler's current.

Despite this discouraging beginning, we went ahead and conducted a conventional search pattern off South Bimini. For five days we dove in an extensive search for the column, covering an area about 1/8 mile by 1 mile.

I did decide to follow up Karen's visual impression of energy radiating up and down from several locations within two circles off South Bimini, each about 3/4 mile in diameter. Joan did a reading and suggested we concentrate on one side within the southernmost of Karen's two circles. Her reading instructed us to begin from Sunshine Inn and then move up the beach to a twisted tree, our onshore marker. We were then to go due west to 30 feet in depth, where lobster traps would be found, and then a shipwreck. The energy source, according to Joan, was located north of the wreck.

I paced the required distance north of the Inn, found the tree described in her reading, then rowed back to the

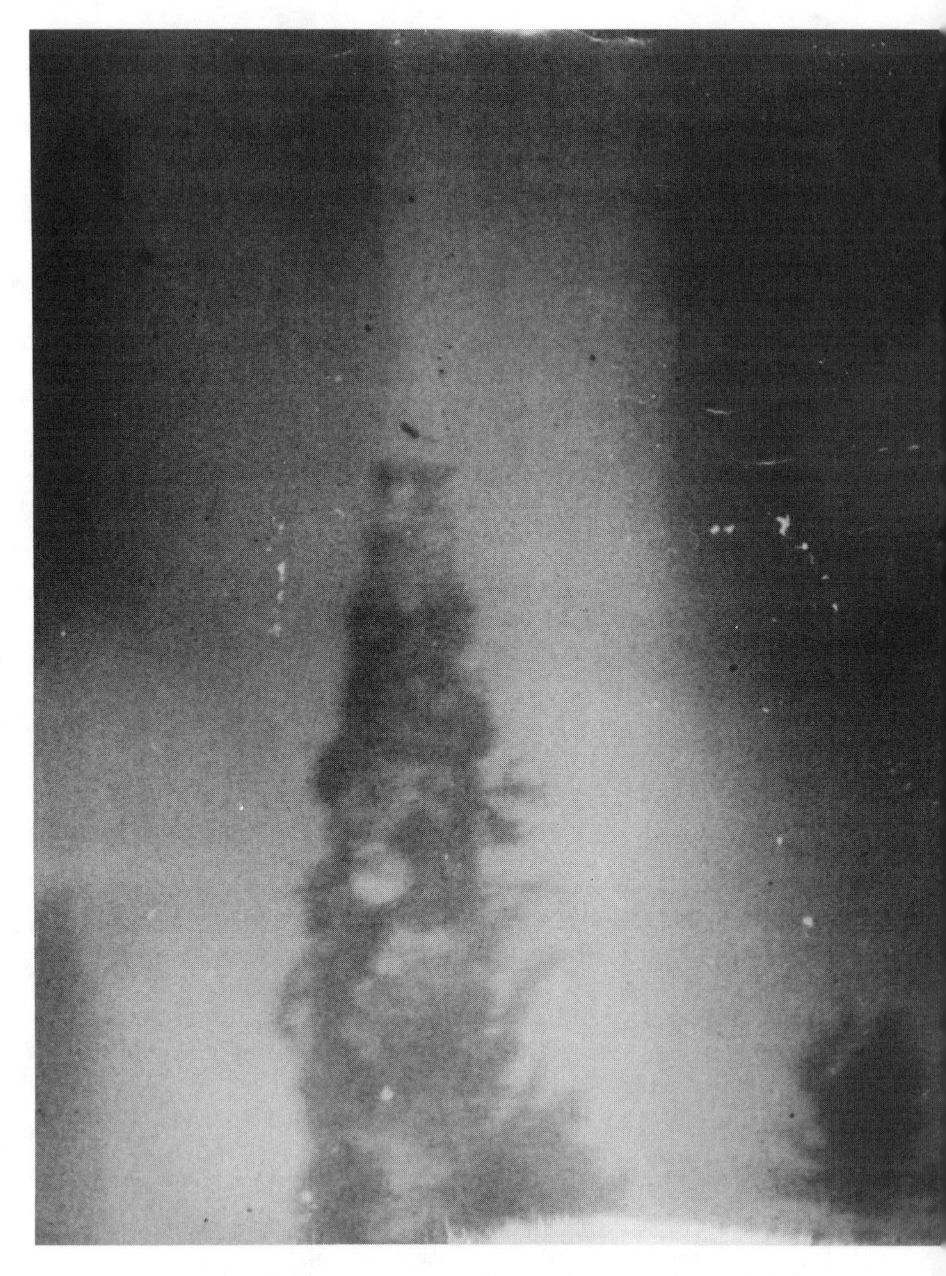

The upper portion of the column. (Courtesy Dr. William Bell.)

The base of the column. (Courtesy Dr. William Bell.)

boat and motored the distance indicated. Sure enough, lobster traps were visible and soon we found on the bottom an arrowlike area of bare sand. Was this the shipwreck? When Joe Libbey, Gary Varney, and I returned several days later, we proceeded to probe the sea bottom with a 10-foot pointed steel rod, to find out just what had caused the unnatural pattern. At depths ranging from 1.3–1.9 meters (about 52–73 inches) we made contact with what felt like both wood and metal. Apparently we had found our wreck!

With all of these elements of Joan's reading checking out, the probability of the column's being in the immediate vicinity seemed high. Meanwhile, intent on our investigation, we nearly failed to observe an unfriendly barracuda. Joe and I became aware that he was quite a huge specimen—between 5 and 6 feet long. His behavior was extremely aggressive, and once Joe had to feint at him with his long steel rod. The barracuda ducked out of range, then turned back to lunge at us and snapped at the rod. He doggedly continued to circle us until we left, later following us back to the boat. The column remained hidden, and it now began to appear that relocating it may prove to be a major engineering task.

Later, I spoke again with Dr. Bell and this time received new information about the column. Joan's reading before the expedition had included the statement that the column was 50–60 feet tall, with most of its length buried in the mud. During our talk, Dr. Bell told me he had also gotten the same height in a description by another psychic. I was grateful to find once again some confirmation for Joan's work.

Although there have been several subsequent attempts to retrieve the column, it seems to have eluded its searchers. And, in a sense, I am relieved not to have found it.

15

IN SEARCH OF A LOST FOUNTAIN

To my knowledge, at least a half dozen psychics have claimed that part of Bimini's ancient past included an extensive system of aqueducts, reservoirs, and healing springs. But it was more than psychic evidence that put me on this trail. In the spring of 1977 Ray McAllister had told me of an amazing observation by "Tex" Treadwell, a retired Navy captain who, while flying a helicopter off Cat Cay, had observed an enormous quantity of fresh water welling up in the sea, far beyond what rainfall in the islands could possibly have produced. It was, he said, as if some unknown water-bearing layer were connecting the Bahamas deep in the earth with mountains farther north in the United States.

There is a long history of tales of a miraculous healing spring at Bimini. Explorer Ponce de Leon searched in vain for Bimini and the Fountain of Youth. (See appendix G.) In the 1920s, Edgar Cayce predicted the discovery of healing springs or wells on Bimini. A few years later, Melaney Freeman, one of the earliest licensed woman pilots, claimed to have found it. Unfortunately a subsequent hurricane must have buried it in mud, as it has not been relocated. More recently, in 1975, the Atlantis scholar, Egerton Sykes, theorized that an ancient healing temple had been situated at Bimini. All of this is perfectly consistent with the general functions of ancient sacred sites as I have been developing them in the present work.

I asked Joan for any information she might be able to receive on such a well or spring. Her readings led me to an underwater spring at the bottom of 40 feet of water off the west coast of South Bimini. I took a sample and a gamma counter showed the water proved to be slightly radioactive. This spurred us on to visit the site of a spring Dick Wingate claimed to have healing properties—a spring that had been pinpointed on a Bimini chart. Later, Jim Richardson, who had flown Dr. Valentine over the banks for a number of years, spotted a light green rectangle in the darker green of the mangrove swamps in the eastern part of North Bimini Island. Dick Wingate told us that Richardson swam in this water for some hours on different occasions and as a result, his arthritis improved so greatly that he is able to play tennis once again. Dick asked for our assistance in continuing his exploration of the spring.

One morning six of us, including Joan and I, John Steele, and Gary Varney, loaded food, machetes, mosquito repellent, scuba gear, and a gamma counter into a Zodiac rubber boat that our friend Dick Hart offered us for the day. At high tide Dick towed the Zodiac over the tidal flats to the east side of North Bimini. Here he quickly left us, as the tide was beginning to ebb fast, exposing the base of mangrove-covered islands and shallow flats. Wingate led us along an 85-yard path he'd begun to clear through the tangled mangrove roots. We picked our way carefully through the knee-deep water to avoid tripping over the sharp stubs of severed roots. Besides motoring we all walked and pulled the boat for several hours across the flats.

The spring was a fascinating experience for all of us. My gamma counter recorded no radioactivity from the water, which Wingate believed to contain radon gas, but the spring did have a high sulfur content. Besides its smell, the water blackened the sterling silver jewelry worn by some members of the expedition. In the course of several hours of swimming in the spring, subtle effects emerged. We all experienced a feeling of euphoria—a gentle high. Furthermore, those with arthritic symptoms lost them during the afternoon. This relief was to last for a matter of days for some, longer for others.

Wingate was very curious to see if probing with iron rods might locate any stone structures such as a retaining wall. The only stone we found was on the bottom, which had a uniform depth of about 8 feet. Diving with scuba gear, Mark and I worked blindly in the dark waters, each clearing away about 18 inches of mud from the stone floor for several feet horizontally. When we compared our experiences afterwards, we agreed that we had felt no evidence of masonry joints; instead we both had touched sea shells set in what we took to be a coral or limestone matrix. In short, we found no evidence of anyone's having improved the spring. On the other hand, its healing effects on our group, both psychological and physiological, were undeniable.

The final afternoon of the expedition, we were anchored over the Road site. Several of the divers wanted to take a last swim in various locations that had earlier seemed promising. I was on deck when Gary Varney swam alongside and told me that he thought he might have something. I quickly donned faceplate and fins and dove in to join him.

Over one edge of the Road he pointed beneath us to what seemed to be a *metate,* a bowl for grinding corn. I dove down to examine it more closely, then swam back to the boat for my underwater tape. Meanwhile, Gary photographed the object. It appeared to be a flat, rectangular shape about 24 inches long by approximately 18 inches wide. In the middle was an oval depression about 3 inches deep and approximately 12 inches on its long axis. Unlike the head and the building block, this object was solidly cemented into the marine limestone of the sea bed, so removal of this discovery would probably carry a high engineering price tag. But at least no casual diver was likely to carry it off.

Because it was (and remains) cemented into the seabed, speculation about its age was possible, even without removing it. When John Gifford's National Geographic Society–sponsored expedition took a core sample of the marine limestone of the seabed adjacent to the Road, he got a uranium-thorium date of about 15,000 B.P.—presumably the age of a marine organism within the core. This would mean that the *latest* date for the *metate* (assuming that it is indeed an artifact) would be 15,000 B.P. Due to the Ice Age, however, the Atlantic at that date would have been about 395 feet lower, leaving the site about 375 feet above the sea. How could the cementing action of seawater supersaturated with an abundance of calcium carbonate have occurred with the site above sea level? Until more is known about this possible artifact, I am inclined to assign it a date somewhat more recent than 8000 B.P.

This new find, apparently cemented into the marine limestone, reminded me of another clue I had been given the previous winter. In 1976 I had met Dr. Marcel Vogel, senior scientist at IBM in San Jose, California, inventor of IBM's Disk Memory, a wizard with plants, and a fine psychic. In February Dr. Vogel had called me to say that a group of psychics he knew had received the impression that the seafloor at the Road had been cemented over since the construction of the site. By drilling cores of some depth, other structures might be revealed.

After finishing the twelve cores we had taken for geological purposes, I had planned to attempt a core five feet into the seafloor. Unfortunately, the limestone blocks that we had cored—considerably harder than ordinary beach rock—had dulled the diamonds on the $1,100 core-and-bit assembly, and we had to postpone this potentially revealing coring project.

As we turned in the chartered ketch in Miami, we made reluctant goodbyes and our group again scattered over the country. As I thought back over the weeks of work, I was satisfied on two points: (1) we'd done what we could with what we had to work with, and (2) we had uncovered some important clues for further investigation at Bimini.

I was ready to sit down with all the evidence and see what conclusions could be drawn about Bimini's prehistory. Only then would I be able to leave this fascinating but maddening part of the earth and turn my attention to other sites that had begun to attract my interest . . .

16

POSEIDIA '78 AND '79

In January of 1978 we motored out of Ft. Lauderdale with Dr. Harold Edgerton of MIT and his 417 pounds of electronic gear—a side-scan sonar and sub-bottom profiler. He had caught the last flight out of Boston's Logan Airport before a major snowstorm closed it down. Although he had worked with the *National Geographic* for many years, for some reason they were not interested in this particular project and so he paid his own expenses. The vessel for Poseidia '78 was a Hatteras 42 diesel cruiser, *Margeo IV,* captained by George Doyle and assisted by the owner, George V. Hersch, Jr. It was chartered by Joseph Barker of Abingdon, Virginia. Other members of the crew included Larry Haley, a diver and reporter from the *Palm Beach Times,* and myself.

Dr. Harold Edgerton of MIT reading his side-scan sonar strip chart aboard Margeo IV.

The main objective was an underwater electronic survey offshore from North and South Bimini Islands. The previous expedition in 1977 had put special attention on the relocation of the so-called lighted column found in 1957 by Dr. Bell. In 1977 the area where it was last seen appeared to be silted over with coral sand. I hoped that the side-scan sonar might help us to relocate the column, which we could then probe with the sub-bottom profiler.

Unfortunately, at the last known location of the column no significant targets were located. Furthermore, the low power and frequencies (6 and 12 kilohertz) barely penetrated the coral sands. Dr. Edgerton's opinion was that a lower frequency and greater power (as used in regular seismographic gear) would be required to get any conclusive scan of the area. Of course, this would require a larger vessel with an adequate generator to support the gear.

We also ran a side-scan sonar profile for about 4 miles on a course line roughly parallel to North Bimini from Entrance Point to near what was then the Standard Properties boundary. Many targets worthy of investigation by divers were found on this run. Numerous rectilinear shapes were recorded on the strip chart; these have yet to be investigated.

At the Road site, the side-scan sonar recorded an oval tangent to the north end of the 600-meter seaward run of the megalithic blocks. The long diameter of the oval stretches 25 meters. This target was first seen in a 1974 aerial photograph, but subsequent underwater searches have not located it. The sonar will detect rises on the bottom not evident to divers; therefore, the oval is probably a structure now covered with marine limestone. Test borings in 1977 indicated a relatively thin surface crust of marine limestone (2–4 inches) which covered loosely consolidated sediments. The evidence in hand suggests the likelihood of uncovering the oval structure

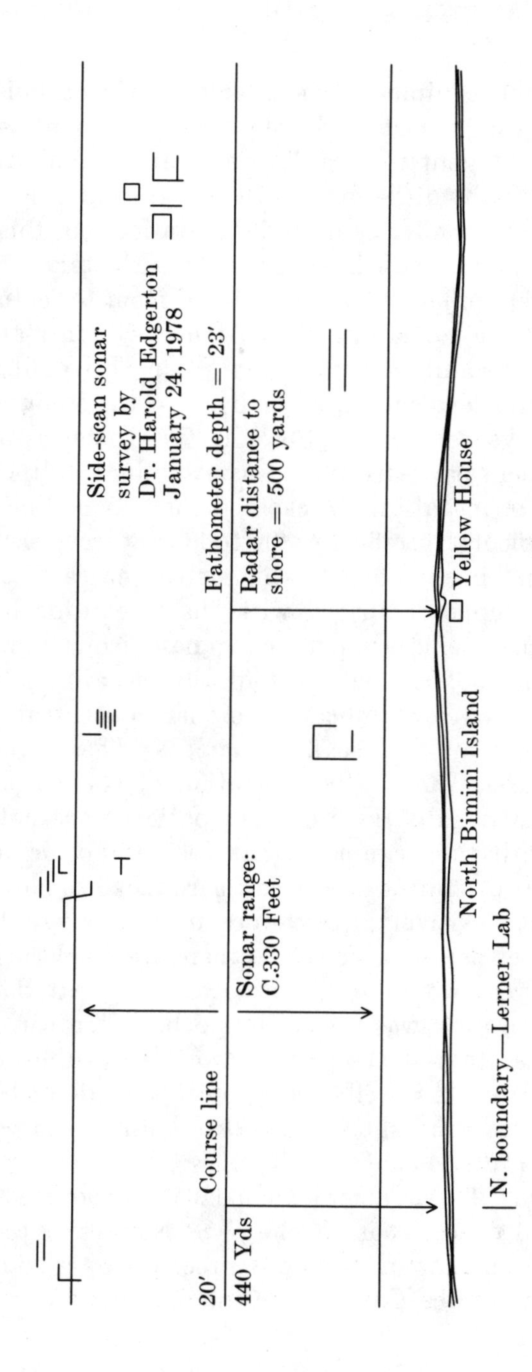

Targets recorded to the west of North Bimini Island by Dr. Edgerton's side-scan sonar (a tracing).

once suitable equipment is available. The location is now known within a meter. Since nature does not usually put ovals tangent to straight-line features, this potential discovery will be yet another argument for extensive human involvement and habitation on this site prior to what we now know as recorded history.

After returning to Virginia Beach from Poseidia '78, I received the lab results from Alan P. Curtner at Virginia Polytechnic Institute and State University. He had done nuclear activation analysis of the twelve cores taken the year before in Poseidia '77. In terms of the trace-element patterns, there was a significant difference between the megalithic blocks and the marine limestone of the seafloor. Specifically, the following trace elements were found in significantly higher concentrations in a seafloor sample, compared with the megalithic blocks: aluminum, arsenic, copper, manganese, samarium, and vanadium. In addition, most of the megalithic blocks cored revealed a significantly higher concentration of strontium than the seafloor sample. These findings argue against an *in situ* formation of the megalithic blocks; that is, the trace element patterns are not consistent with the argument that the megalithic blocks were naturally formed in place where they are now seen. To put it positively, they were more likely cut and shaped elsewhere and moved to their present location.

As the year wound down, it began to appear that the Poseidia project was beginning to be taken seriously. The first edition of the present work was published by Prentice-Hall. As well, the project was described in *Scientific American*'s book, *In the Spirit of Enterprise: From the Rolex Awards.*

Poseidia '79 was a major expedition conducted with the official cooperation of the U.S. Navy and the Defense Mapping Agency. It was preceded by a slide lecture given at the Explorers Club in New York City.

Afterward, Lt. Paul Evancoe also explained the objectives of the coming expedition. The presentation elicited many thoughtful questions from the audience, many of whom were from various disciplines of the earth sciences.

The expedition was funded by the Max and Victoria Dreyfus Foundation of New York City. The 42-foot ketch, *Lark,* was chartered out of Miami. The vessel had ample deck space for a scuba compressor and other useful gear on loan from the Navy. Of course, as with all chartered vessels, it presented a few maintenance challenges, but that is the name of the game. We assembled in Miami on June 4.

I captained the *Lark* and Paul Evancoe was the chief executive officer. Terry Mahlman was the science officer and staff archaeologist; Janet MacArthur the medical officer. The diving team included Charles E. Donnally, Rob Peterman, Stan Janecka, and Roger Crossland.

The plane carrying some of the team from Virginia was late arriving and so we did not cast off from Monty Trainer's Marina until 7:50 P.M. Considering the draft of the vessel, I took a calculated risk on an ebbing tide and a moonless night, tried to run the channel south of Key Biscayne to Fowey Rock Lighthouse. As often happens, we spent nearly seven hours aground during the night. In the morning on a flood tide we reached Fowey Rock at about 7 A.M. and took our departure for Bimini. Despite the time aground, we were at sea much earlier than we would have been had we waited for a morning departure from the marina. The week ahead promised to be quite busy, particularly because of the need for the Navy people to return on June 11.

We made good time crossing the Gulf Stream and docked at Alice Town at 1:15 P.M. with cloudless skies and the temperature at 100 degrees. I was able to clear customs and immigration very rapidly and once again we were underway, this time to the Road site, which we reached at 3:30 P.M. and dropped anchor. We did an initial reconnaissance of the site so that everyone would have a feel for the project. Being SEALS and UDT personnel, the divers immediately came to regard the operation as a holiday. Before it was over, however, they would have a few small challenges.

On Wednesday, June 6, the group surveyed the Road site with an extremely sensitive metal detector (sensitive to ferrous and nonferrous metals). For our purposes, it was the equivalent of a magnetometer. At several locations within the J metal returns were noticed, although there was no metal on the surface. This meant that whatever was being sensed was under 2–4 inches of marine limestone covering the seabed. Checking out this finding would, of course, require a permit from the Bahamian government. In the afternoon, the dive team of Paul Evancoe and Rob Peterman found a piece of worked stone which was logged as the "keystone." This was the fourth artifact found during the Poseidia project.

The next morning I dove on this find with Terry Mahlman. As I sketched it out on my underwater slate, with its measurements, my amazement grew. Every dimension was divisible by 10 centimeters! I measured it again and again—only to find the same measurements. How would we ever discover the reason for this astonishing correlation with the metric system—much less the function and culture of this artifact?

That afternoon, to add to the drama, the divers were doing an underwater recon of the Paradise Point and Crossing Rocks area when they found an underwater tunnel that penetrated the inner Crossing Rock, a coral head. On the last dive, Paul Evancoe found a piece of worked stone about 29 centimeters by about 12.5 centimeters. As is evident in the illustration, its shape was clearly human-made. All the more exciting was the fact that it was found in an underwater cave. The skeptics are fond of saying that each of the artifacts found during the Poseidia project fell off a passing ship—that is, they could not be used to defend human intervention on this site. This was the fifth artifact found during the Poseidia project.

Since the expedition, C. Lavett Smith, chairman and curator of the department of ichthyology at the American Museum of Natural History, showed the artifact to two colleagues, Dr. Gordon Ekholm and Dr. Junius Bird, curators emeritus. Both are distinguished in Meso-American archaeology. They both agreed that it had been shaped by human intervention, but could not identify its function or relate it to any known pre-Columbian culture.

The last three days of Poseidia '79 were also very productive. On June 8, a survey by divers in an area 200 by 1,600 meters found nothing down the west coast of North Bimini in the area where Dr. Edgerton's side-scan sonar had logged many rectilinear shapes. In the afternoon, however, Paul Evancoe and Stan Janecka, while deploying the very sophisticated temperature-sensor device they had brought along, detected heat off South Bimini. This was during a search of the area where the "lighted column" had been found by Dr. Bell.

In terms of its implications, this discovery was worth the whole trip. It was so important that despite choppy sea conditions and wind, on Saturday morning, June 9, 1979, we anchored the *Lark* near the site and made several dives. On the final dive, Paul Evancoe and Stan Janecka found a geothermal anomaly that has yet to be explained. Due to a 2-knot current and a 4- to 6-foot chop, when they surfaced they found it necessary to fire a smoke flare to give us their position. Furthermore, the tide had carried them downstream about 200 yards from

POSEIDIA '79
Preliminary Sketches of
Two Artifacts

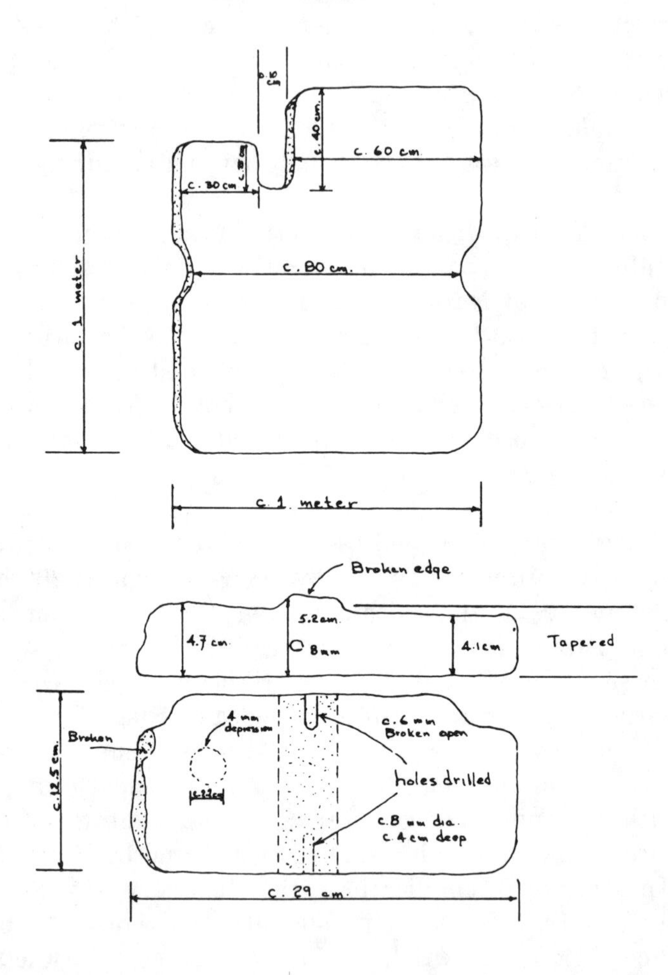

Two artifacts discovered during Poseidia '79: the "keystone," whose dimensions are divisible by 10cm; and the piece of worked stone approximately 29cm long with drill holes and other evidence of human handiwork. It relates to no known culture and geologically dates to the Permian (250–290 million years ago). It was discovered in a cave near the main site by Paul Evancoe.

the *Lark*'s anchorage. We quickly got underway and picked them up.

Once aboard, they told of a pool of warm water that rose at least 10 feet above the seafloor and extended about 300 feet across. The temperature was about 100 degrees Fahrenheit. The marine life around it suggested that it had been in existence for some time, despite the constant tidal currents in the vicinity in 60 feet of water. The presence of water of this temperature, of course, rendered our extremely sensitive heat sensor useless. Because of the sea conditions, further diving was unsafe and so the crew was given liberty in port.

On June 10, we interviewed an Australian named Gene on his trimaran at anchor. The divers had met him the previous evening in Alice Town, where he had told them of a discovery of a ruined standing-stone circle similar to the inner circle of Stonehenge. He had found it thirteen years earlier in about 30 feet of water between North Rock Light and the north tip of North Bimini Island. He sketched it out for us and said that he had not seen it for about nine years. A lead like this could not be ignored. He offered to go with us and so we spent the afternoon towing two teams of divers alternatively to survey about 4 square miles. Unfortunately, the results of this effort were negative. After the expedition a Navy aerial photo was consulted that seemed to show the location of this site at the edge of the area we had surveyed. If the site exists, it could revolutionize New World archaeology.

After this last search north of North Bimini, our project was concluded, since some of the Navy people had to get back to duty. We made it back to Miami that evening, cleaned up the vessel the next morning and went our separate ways.

I then flew to Nassau to explain the expedition's results to Senator Doris L. Johnson, president of the

Bahamian Senate and director of the Bahamian Antiquities Institute (of which I was then the honorary director of research). It was also necessary to get permission from the Bahamian Ministry of Agriculture, Fisheries, and Local Government to remove the small, worked stone from the Bahamas for archaeological opinions. A tropical depression over Nassau caused two days to be lost due to canceled flights.

After the expedition, preliminary discussions with Navy oceanographers suggested that what we had observed off South Bimini was geothermal activity produced by superheated gases coming from tectonic or seismic activity deep within the earth. Furthermore, we were told, such geothermal activity (until our find) was unknown north of a line running from Cuba to the Yucatan.

Still later, Jabe Breland, a marine chemist at the University of South Florida, informed us of several warm-water springs in the Gulf of Mexico. Thus our investigations appear to have added to geophysical data in the region.

The geothermal activity found during Poseidia '79 is consistent with a hypothesis of healing springs (or a "fountain of youth") first suggested by the indigenous Lucayan Indians of the Bahamas in interviews with the Spanish shortly after the arrival of Columbus. On the other hand, such geothermal activity is *not* consistent with the usual geological picture of the Bahamas. Geologically, the archipelago is taken to be stable, not dynamic. Geothermal phenomena are usually associated with seismic and tectonic activity, not a stable geology.

Furthermore, this more dynamic geology is in accord with a 60-kilometer fault line discovered in a Skylab photo, a fault running north and south on the Great Bahama Bank with a 12- to 15-foot vertical displacement (according to a fathometer profile taken on a U.S.

Defense Mapping Agency field trip). Previous Poseidia expeditions, as noted earlier, had investigated a freshwater spring on East Bimini that seems to have some healing properties *and* a freshwater spring off South Bimini that has some radioactivity (as detected by a gamma counter). This latter spring was within 500 yards of the geothermal activity discovered during Poseidia '79.

As usual, this expedition identified more problems to solve than it found answers for. Ten years later, in 1989, Dimitri Rebikoff would remark in reference to the prehistory of the Bahamas, "There is a hundred years of work out here."

17

WHAT MAY IT ALL MEAN?

At this stage of a project that may ultimately require a decade for its completion, it is impossible to be definitive. Yet enough has been learned to convince me that Bimini may ultimately be recognized as the major archaeological discovery of this century in the New World. Perhaps we are fortunate that the ancient past of Bimini is essentially a blank for historians. We are not saddled with elaborate (and possibly erroneous) preconceptions. We are also fortunate in that the discoveries of science on earth (and in space explorations as well) have so expanded our consciousness that we are more open to new revelations about the prehistory of humankind.

As should be clear by now, however, the underwater sites around Bimini present problems far more difficult to solve than more recent sites on land. Months after the end of the Poseidia '77 expedition, I sat down with notebooks before me and decided to take the plunge. Now I must put aside my hesitations and let the evidence of the past few years sort itself into a meaningful whole. I decided to let the chips fall where they may—it was too late for fence-sitting. My conclusions, based on our explorations at Bimini, finally emerged as follows:

1. Bimini was occupied long before the Lucayan inhabitants found by Columbus.
2. Whatever the precise geological history of the limestone blocks off Paradise Point, their order and patterning clearly suggest human intervention.
3. Dr. Valentine's original claim—that the Road was the remains of a megalithic structure—seems valid.
4. Furthermore, our survey uncovered remnants of a sacred geometry on the Road. Despite the local magnetic anomalies and the unresolved question of possible astronomical orientations, the internal geometry associated with the chockstone, the arrow, and the obelisks will certainly stand up in the face of close scrutiny—as will the numerical patterns.
5. The site was definitely not a road, but one of its functions may have included the utilitarian one of protecting a temple from surrounding waters.
6. During the Poseidia project a total of five pieces of worked stone were discovered, none of which is related to any known pre-Columbian culture.
7. Natural springs with a healing potential are indicated as having been a part of a large sacred complex.
8. As recently as six to seven thousand years ago,

the Bimini complex would have been located near the northwestern extremity of a land mass as long as the present state of Florida and wider.

9. With the help of our psychic assistants, Dr. Bell's incredible column with its stunning implications has become a more tangible objective.

Because Bimini's underwater sites present problems far more difficult to solve than more recent sites discovered on land, my present reconstruction of the events of Bimini's prehistory is based on dates from paranormal channels as well as those from more conventional scientific authorities and methods. Some might find paranormal sources unacceptable; yet for me, when these dates cluster with dates from other sources (and also with each other), their probable validity increases. I suspect that the cementing of the seafloor under the megalithic blocks is comparatively recent and, in fact, now covers structures yet undiscovered.

The prehistoric scenario for Bimini may have gone something like this: Based on Plato's contention that Atlantis was in the Atlantic, which I am inclined to support, we find Atlantis somewhere in size between a large island and a small continent. During its long existence, many colonies were established around the Atlantic basin; perhaps even one on Santorini, which was then that island's *original* culture.

As for the date of Atlantis' origin, most paranormal sources place it well before the following probable geophysical and geological events: Eighty thousand years ago, polar ice was at a minimum, as evidenced by high sea levels determined from coral terraces in Barbados, New Guinea, and Hawaii (Hays, *et al*). This was the time of the warmest climate in the last phase of the Upper Pleistocene, just before the ice began to advance once again. The Atlantean lands could well have had the climate reflected by legends of the Garden of Eden. At this time, Neanderthals were dominant in Europe, and assuming no vertical displacement of the island, Bimini would have been completely under water.

The onset of the next major glaciation seems to coincide in time with a shift of the earth's north magnetic pole from the Yukon to the Greenland Sea. This, according to Hapgood, took place from 78,000 B.C. to 73,000 B.C. (80,000–75,000 B.P.) Thus the earth's core (whose outer portion is the source of the earth's inner magnetic field) was rotating at a different speed than that of the crust—one slipping over the other. Hapgood feels that the evidence suggests widespread volcanism at the time of this slipping. This may have also been a time when the usually slow horizontal movement of the continental plates was briefly speeded up. Such geophysical conditions would have been overwhelming. Because the evidence also suggests that the ice ages recur periodically, timed by subtle changes in the earth's orbit around the sun (Hays, *et al*), such violence has likely been periodic in the earth's history.

Velikovsky challenged the traditional geological theory of a slow advance of ice with his proposed astrophysical events—including a near passage of Venus. Perhaps his scenario is descriptive of several chapters in the earth's prehistory; but probability seems to be on the side of periodic ice ages *initiated* and *terminated* by minor changes in the earth's orbit and accompanied (at their beginning and end) by violent geophysical and geological events.

From 53,000 B.C. to 48,000 B.C. the earth's north magnetic pole shifted from the Greenland Sea to Hudson's Bay, according to Hapgood. Mute testimony to the violence on the planet's surface at this time was found in Siberia in 1901 with the discovery of a still frozen carcass of the mammoth at Beresovka. This animal died between 45,500 B.C. and 37,000 B.C. (a carbon-14 date) while grazing on buttercups and other summer vegetation then found in its habitat. Its body was immediately quick-frozen, never to thaw until 1901. Only the most stupendous geological and meteorological events could have brought this about. Considering that the carbon-14 date for the mammoth's death is approximate, its demise could actually have been closer to the end of the pole's drift—48,000 B.C.

According to the Edgar Cayce material, a world conference was held on Atlantis in 50,727 B.C. to deal with the threat of large animals. Sometime afterward, while attempting to kill them with energy from the crystal, the Atlanteans triggered earth changes that reduced Atlantis to five islands. According to Cayce, Bimini is the present remains of one of these. I am inclined to regard Bimini as an Atlantean colonial site or the location of a different culture parallel in time to Atlantis. We have already seen hints that this was probably a time of vast natural changes on the earth's surface. Even today, the Atlantic basin gives evidence of considerable geological instability. If the possibility of such a powerful ancient technology is granted, then earth changes in progress might have been escalated by human misuse of natural forces. Furthermore, if the relationship between magnetic-pole shifts and earth changes theorized by Hapgood is valid, then the earth changes Cayce cited probably took place sometime after 48,000 B.C.

From the evidence of paleomagnetism, (a study of the earth and its magnetic fields), 28,000 B.C. was the time of a geomagnetic pole reversal. For 8,000–10,000 years, worldwide volcanic activity had been fairly dormant, according to Hapgood. Volcanism resumed around 27,800 B.C. According to Joan, 28,000 B.C. was the time of a major earth change, including a 7-degree rotation of the plate upon which the Bahamas ride. Cayce gave this same date as that of a second violent earth change, leading to migrations from Atlantis and the establishment of colonies on the southeastern coast of the present United States, in the Yucatan in Mexico, Brazil, Peru, Spain (the Basque people), England, and Ireland.

The natural catastrophes of this era may also have resulted in a cultural regression of some escaping Atlanteans to the various Cro-Magnon cultures. (Lewis Spence claimed that the Atlantean migration was the basis for Cro-Magnon culture.) At this time the Bimini Road was 15 feet above water (Milliman and Emery). In the midst of these awesome events, the Pleions arrived at Bimini to begin their mission. (Carol had dated this event at 28,000 B.C.) To raise human consciousness and heal physical ills, they proceeded to build temples constructed with what John Michell calls a sacred geometry, in which were used certain sacred numbers relating to the structure of the universe.

Five thousand years after the Pleions arrived at Bimini (or 23,000 B.C.), the unstable earth's crust caused such a forceful explosion of the volcano at Santorini that the ashes from this explosion blew across the Mediterranean to Italy.

According to the Cayce material, there was a destruction of Atlantean land in the region of the Sargasso Sea (the Atlantic off the Bahamas, roughly) in 18,200 B.C., 7,500 years before the final destruction. Two hundred years later (18,000 B.C.) the earth's magnetic field returned to its normal polarity (Cox). Then, from 15,000 B.C. to 10,000 B.C., the north magnetic pole moved from Hudson's Bay to its present location (Hapgood). As earlier suggested, this phenomenon may be related to cataclysmic earth changes, having created increasing instability in the earth's crust.

The following events, because of their complexity, can be most clearly grasped in tabular form beginning on the next page:

10,700 B.C. Final date of destruction of Atlantis, according to Cayce. He also stated that Bimini was a part of Poseidia, one of the two islands of Atlantis remaining after 28,000 B.C. Carol's readings suggest Atlantean refugees began arriving at Bimini and other locations in the New World.

10,350 B.C. End of another reversal of the earth's magnetic field: the "Gothenburg magnetic flip." This reversal may be related to the final shift of the magnetic pole claimed by Hapgood between 15,000 and 10,000 B.C. and is very close to Cayce's date.

9600 B.C. Emiliani's date of a worldwide flood (from oxygen isotopes in Foraminifera in cores from the Gulf of Mexico). The Pleistocene Age ends with a rapid melting of the ice in high latitudes.

9570 B.C. Plato's date for the end of Atlantis. The difference between Cayce and Plato—about 1,100 years—may indicate that, instead of only one, there were two violent earth changes.

6031 B.C. According to Carol, a cataclysm destroys the site at Bimini. About this time, the Road would have been either 32 feet above the sea (date of beach rock now 50 feet under the sea at Cat Cay, 9 miles south of Bimini: Kornicker) or 80 feet above the sea (for Atlantic sea levels, see Milliman and Emery).

4021 B.C. The culture associated with the Bimini site abandons the area (Carol). Between this date and the eventual occupation by

> at least one circum-Caribbean culture, the Lucayans, there may have been other transatlantic influences, including Egyptian, Phoenician, and Celtic. The latter two have also left their inscriptions carved in megalithic stones at Mystery Hill in New Hampshire. Cyrus Gordon claims continued transatlantic voyaging by Bronze Age sea lords from about 3000 B.C. to about 1200 B.C. Egerton Sykes sees Bimini as having been occupied down into historical times.

There is mounting evidence that we may be approaching another period of great instability for the earth's surface. The strength of the geomagnetic field appears to be decreasing as we move toward the next reversal of the earth's magnetic poles, which Harwood and Malin predict for A.D. 2030. Past field reversals of the planet have been correlated with the extinction of various life forms, possibly a result of increased cosmic radiation. Hapgood's research has amassed considerable evidence of violent beginnings and endings of the ice ages. However it works, this magnetic sequence also seems to be related to periods of earth changes, including the onset of the ice ages and possibly accelerated earth-plate movement, volcanism, and earthquakes. (The latter activity has increased enough in recent years that much effort is now being made to improve earthquake prediction.)

All of these events may have some relation to changes in the earth's orbit around the sun. Such periodic natural changes have been used to explain the timing of the ice ages and when, during the next several thousand years, the ice ages may be expected to return (Hays, *et al*). These major climatic changes have thus apparently occurred as a result of regular, long-term changes in the obliquity of the earth's axis (over a period of about 41,000 years) and the precession of the equinoxes (a period of about 26,000 years). These long-term influences apparently produce periodic eras of instability for the earth's crust.

If the earth has regularly experienced such extreme conditions of instability—unrecorded by history for obvious reasons—and if previous civilizations have been able to produce a minimal effect upon their environment through the harnessing of natural forces, couldn't they, by some miscalculation in dealing with an energy source, also trigger the mechanisms of destruction?

18

AFTERWORD

In June of 1975, in an issue of *Atlantis* devoted to the Bermuda Triangle mystery, Egerton Sykes described Bimini as the site of a great religious complex occupied down into historical times. He called Bimini a "harbor" and an "official building, probably a temple." From his studies of the ancients he suspected that the temple was in fact the Murias of the Tuatha De Danann, the legendary pre-Celtic tribe long important in the Irish pantheon. Sykes went on to claim that the crystal skull found on an expedition conducted by F. A. Mitchell-Hedges and described in Richard Garvin's book, *The Crystal Skull,* was originally located at Bimini.

Anna Le Guillon, Mitchell-Hedges' adopted daughter, found the skull in 1927 at Lubaantun in former British Honduras, now Belize. Garvin's account of the skull reveals it to be an extraordinary remnant of a high state of scientific knowledge that included crystallography. Those who have investigated the skull report a variety of paranormal phenomena associated with it. Sykes also linked the Bimini site with what he calls the original concept of the Holy Grail, a healing cup. Murias, he further says, was the Temple of the Translucent Walls and the Golden Gates. The healing temple, he said, was dedicated to the god Min—hence the name of Bimini—and the Benu bird, both divinities of rejuvenation (see appendix G). The temple's healing function led to a proliferation of legends that ultimately caused Ponce de Leon and countless others to seek the secret of eternal youth; as Sykes put it, "Ponce de Leon came seeking rejuvenescence and landed at Miami instead of Bimini." Sykes further maintained that the temple had a series of occupations by various cultures including the Egyptians and later even the Phoenicians. Declining to identify his sources, he also claimed that an Egyptian temple to Isis was once located in the Caribbean—not at Bimini.

It may be many years before these claims can be fairly tested. Two of his ideas about Bimini, however, ultimately proved to parallel information from psychic readings at the site. Two years before Sykes' article, a French book had argued that history had begun at Bimini. The thesis of this book, *L'histoire commence à Bimini* by Pierre Carnac, the pen name of a Rumanian now living in Paris, is that megalithic culture diffused from west to east, not the reverse as is usually believed. The author draws upon Celtic accounts of early Atlantic voyages and also reasons from what was known of the Bimini site in 1973 and the Mystery Hill site in New Hampshire. Carnac also gives us an interesting linguistic clue as to the possible origin of the word *Bimini.* Some of the early observers of the culture Columbus found upon his arrival in the Bahamas reported that the natives were very much like the Tainos of Haiti, a tribe closely related to the Arawaks. In 1645, Father Raymond Breton, missionary to the Antilles, prepared a dictionary of the Taino language. Working from it, Carnac translates Bimini as "The Island of the Wreath" (or crown) or "The Island of the Old Wall" (or ruin). These linguistic clues speak of possible visits to Bimini when the site was more clearly an archaeological ruin.

Other than these two authors' speculations and the predictions of Cayce, our investigations took place in what might be called the no-man's-land of history. Should the Bimini site come to be generally recognized as an ancient megalithic site, its discovery would represent an important addition to the New World's prehistory, but it would also contribute greatly to the deepening sense of human antiquity on the planet. This emerging perspective comes at a time when humankind has become intensely interested in its past, the present problems having reached overwhelming proportions, with few answers to these agonizing dilemmas in sight.

I have sought to share the inner development of certain convictions which run counter to the academic traditions about prehistory that were a part of my training. In addition to what I hope is a clear expression of my feelings, I have also tried to suggest the kind of evidence that, for me, supports the Atlantean hypothesis at least as well as it does the present concept of the evolution of civilization from a standing start in the Fertile Crescent 10,000 years ago.

Because of the complexity of the problem, conclusively proving or disproving the former existence of Atlantis is unlikely at the present. In essence, this book advances an argument based on *probability*—a form of reasoning not at all unusual in modern science and technology. Consider, for example, the odds faced in the oil industry, keeping in mind that wells are incredibly expensive to drill. Drilling in the most probable locations, eight out of one hundred wells pay off, and out of these eight, in only two will it be economically feasible to pump.

My investigation of the problems here considered has slowly but surely forced me to accept Plato as an authoritative source of information about a historical fact: that Atlantis *did* exist. Going even further, I believe in this strongly enough to feel that further research and exploration—and the major expenses involved—are justified.

It would appear that now, as never before, any significant expansion of our understanding of prehistory will have a crucial effect upon our present civilization, perhaps even upon its survival.

APPENDICES

APPENDIX A: PLATO'S LEGEND OF ATLANTIS

It is a singular circumstance, that though there is not, perhaps, any thing among the writings of the "antients" that has more generally attracted the attention of the learned in every age than the Atlantic history of Plato, yet no more than one single passage of about twenty or thirty lines has, prior to my translation of the *Timaeus,* appeared in any modern language. Much has been said and written by moderns respecting the Atlantic island; but the extent of the original source has not even been suspected.

That the authenticity of the following history should have been questioned by many of the moderns, is by no means surprising, if we consider that it is the history of an island and people that are asserted to have existed *nine thousand years* prior to Solon; as this contradicts the generally received opinion respecting the antiquity of the world. However, as Plato expressly affirms, that it is a relation in *every respect true,* and, as Crantor, the first interpreter of Plato, asserts, "that the following history was said, by the Egyptian priests of his time, to be still preserved inscribed on pillars," it appears to me to be at least as well attested as any other narration by any ancient historian. Indeed, he who proclaims that "truth is the source of every good both to Gods and men," and the whole of whose works consists in detecting error and exploring certainty, can never be supposed to have willfully deceived mankind by publishing an extravagant romance as matter of fact, with all the precision of historical detail.

Some learned men have endeavored to prove that America is the Atlantic island of Plato; and others have thought that the extreme parts of Africa toward the south and west were regarded by Plato in this narration. These opinions, however, are so obviously erroneous, that the authors of them can hardly be supposed to have read this dialogue, and the first part of the *Timaeus;* for in these it is asserted that this island, in the space of one day and night, was absorbed in the sea.

I only add, that this dialogue is an appendix, as it were, to the *Timaeus,* and that it is not complete, Plato being prevented by death from finishing it, as we are informed by Plutarch in his life of Solon.

CRITIAS

It is requisite that all we shall say should become in a certain respect an imitation and a resemblance. But we see the facility and subtlety which take place in the representation exhibited by pictures of divine and human bodies, in order that they may appear to the spectators to be apt imitations. We likewise see, with respect to the earth, mountains, rivers, woods, all heaven, and the revolving bodies which it contains, that at first we are delighted if any one is able to exhibit but a slender representation to our view; but that afterwards, as not knowing any thing *accurately* about such-like particulars, we neither examine nor blame the pictures, but use an immanifest and fallacious adumbration respecting them. But when any one attempts to represent our familiar animadversion of them, and we become severe judges of him who does not perfectly exhibit all the requisite similitudes. It is likewise necessary to consider the same thing as taking place in discourse. For, with respect to things celestial and divine, we are delighted with assertions concerning them that are but in a small degree adapted to their nature; but we accurately examine things mortal and human . . .

If, then, we can sufficiently remember and relate the narration which was once given by the Egyptian priests, and brought hither by Solon, you know that we shall appear to this theatre to have sufficiently accomplished our part. This, therefore, must now be done, and without any further delay.

But first of all we must recollect, that the period of time from which a war is said to have subsisted between all those that dwelt beyond and within the pillars of Hercules, amounts to nine thousand years: and this war it is now requisite for us to discuss. Of those, therefore, that dwelt within the pillars of Hercules, this city was the leader, and is said to have fought in every battle; but of those beyond the pillars, the kings of the Atlantic island were the leaders. But this island we said was once larger than Libya and Asia, but is now a mass of impervious mud, through concussions of the earth; so that those who are sailing in the vast sea can no longer find a passage from hence thither.

As, therefore, many and mighty deluges happened in that period of nine thousand years (for so many years have elapsed from that to the present time), the defluxions of the earth at these times, and during these calamities, from elevated places, did not, as they are elsewhere wont to do, accumulate any hillock which deserves to be mentioned, but, always flowing in a circle, at length vanished in a profundity. The parts, therefore, that are left at present are but as small islands, if compared with those that existed at that time; and may be said to resemble the bones of a diseased body; such of the earth as was soft and fat being washed away, and a thin body of the country alone remaining.

And these writings were in the possession of my grandfather, and are now in mine: they were likewise the subject of my meditation while I was a boy. But it will require a long discourse to speak from the beginning, as I did before, concerning the allotment of the Gods, and to show how they distributed the whole earth, here into larger, and there into lesser allotments, and procured temples and sacrifices for themselves.

Neptune, indeed, being allotted the Atlantic island, settled his offspring by a mortal woman in a certain part of the island, of the following description. Towards the sea, but in the middle of the island, there was a plain, which is said to have been the most beautiful of all plains, and distinguished by the fertility of the soil. Near this plain, and again in the middle of it, at the distance of fifty stadia, there was a very low mountain. This was inhabited by one of those men who in the beginning sprung from the earth, and whose name was Evenor. This man living with a woman called Leucippe had by her Clites, who was his only daughter. But when the virgin arrived at maturity, and her father and mother were dead, Neptune being captivated with her beauty had connection with her, and enclosed the hill on which she dwelt with spiral streams of water; the sea and the land at the same time alternatively forming about each other lesser and larger zones.

Of these, two were formed by the land, and three by the sea; and these zones, as if made by a turner's wheel, were in all parts equidistant from the middle of the island, so that the hill was inaccessible to men. For at that time there were no ships, and the art of sailing was then unknown. But Neptune, as being a divinity, easily adorned the island in the middle; caused two fountains of water to spring up from under the earth, one cold and the other hot; and likewise bestowed all-various and sufficient aliment from the earth. He also begat and educated five male twins; and having distributed all the Atlantic island into ten parts, he bestowed upon his first-born son his maternal habitation and the surrounding land; this being the largest and best division. He likewise established this son king of the whole island, and made the rest of his sons governors. But he gave to each of them dominion over many people, and an extended tract of land. Besides this, too, he gave all of them names. And his first-born son, indeed, who was king of all the rest, he called Atlas, whence the whole island was at that time denominated Atlantic. But the twin son that was born immediately after Atlas, and who was allotted the extreme parts of the island, towards the pillars of Hercules, as far as to the region which at present from that place is called Gadiric, he denominated according to his native tongue Gadirus, but which we call in Greek Eumelus. Of his second twin offspring, he called one Ampheres, and the other Eudaemon. The first-born of his third offspring he denominated Mneseus, and the second Autochthon. The elder of his fourth issue he called Elasippus, and the younger Mestor. And, lastly, he denominated the first-born of his fifth issue Azaes, and the second Diaprepes. . . .

The race of Atlas was by far the most honourable; and of these, the oldest king always left the kingdom, for many generations, to the eldest of his offspring. These, too, possessed wealth in such abundance as to surpass in this respect all the kings that were prior to them; nor will any that may succeed them easily obtain the like. They had likewise every thing provided for them which both in a city and every other place is sought after as useful for the purposes of life. And they were supplied, indeed, with many things from foreign countries, on account of their extensive empire; but the island afforded them the greater part of every thing of which they stood in need.

In the first place, the island supplied them with such things as are dug out of mines in a solid state, and with such as are melted: and orichalcum, which is now but seldom mentioned, but then was much celebrated, was dug out of the earth in many parts of the island, and was considered as the most honourable of all metals except gold. Whatever, too, the woods afford for builders the island produced in abundance. There were likewise sufficient pastures there for tame and savage animals; together with a prodigious number of elephants. For, there were pastures for all such animals as are fed in lakes and rivers, on mountains and in plains. And, in like manner, there was sufficient aliment for the largest and most voracious kind of animals. Besides this, whatever of odoriferous the earth nourishes at present, whether roots, or grass, or wood, or juices, or gums, flowers, or fruits, these the island produced, and produced them well. Again, the island bore mild and dry fruits, such as we use for food, and of which we make bread (aliment of this kind being denominated by us leguminous), together with such meats, drinks, and ointments, as trees afford. Here, likewise, there were trees, whose fruits are used for the sake of sport and pleasure, and which is difficult to conceal; together with such dainties as are used as the remedies of satiety, and are grateful to the weary. All these an island, which once existed, bore sacred, beautiful, and wonderful, and in infinite abundance.

The inhabitants too, receiving all these from the earth, constructed temples, royal habitations, ports, docks, and all the rest of the region, disposing them in the following manner: In the first place, those who resided about the ancient metropolis united by bridges those zones of the sea which we before mentioned, and made a road both to the external parts and to the royal abode. But the palace of the king was from the first immediately raised in this very habitation of the God and their ancestors. This being adorned by one person after another in continued succession, the latter of each always surpassing the former in the ornaments he bestowed, the palace became at length astonishingly large and beautiful. For they dug a trench as far as to the outermost zone, which commencing from the sea extended three acres in breadth, and fifty stadia in length. And that ships might sail from this sea to that zone as a port, they enlarged its mouth, so that it might be sufficient to receive the largest vessels.

They likewise divided by bridges those zones of the earth which separated the zones of the sea, so that with one three-banked galley they might sail from one zone to the other; and covered the upper part of the zones in such a manner that they might sail under them. For the lips of the zones of earth were higher than the sea. But the greatest of these zones, towards which the sea directed its course, was in breadth three stadia: the next in order was of the same dimension. But, of the other two, the watery circle was in breadth two stadia; and that of earth was again equal to the preceding circle of water; but the zone which ran round the island in the middle was one stadium in breadth.

The island which contained the palace of the king was five stadia in diameter. This, together with the zones, and the bridge which was every way an acre in breadth, they inclosed with a wall of stone, and raised towers and gates on the bridges according to the course of the sea. Stones, too, were dug out from under the island, on all sides of it, and from within and without the zones: some of which were white, others black, and others red: and these stone quarries, on account of the cavity of the

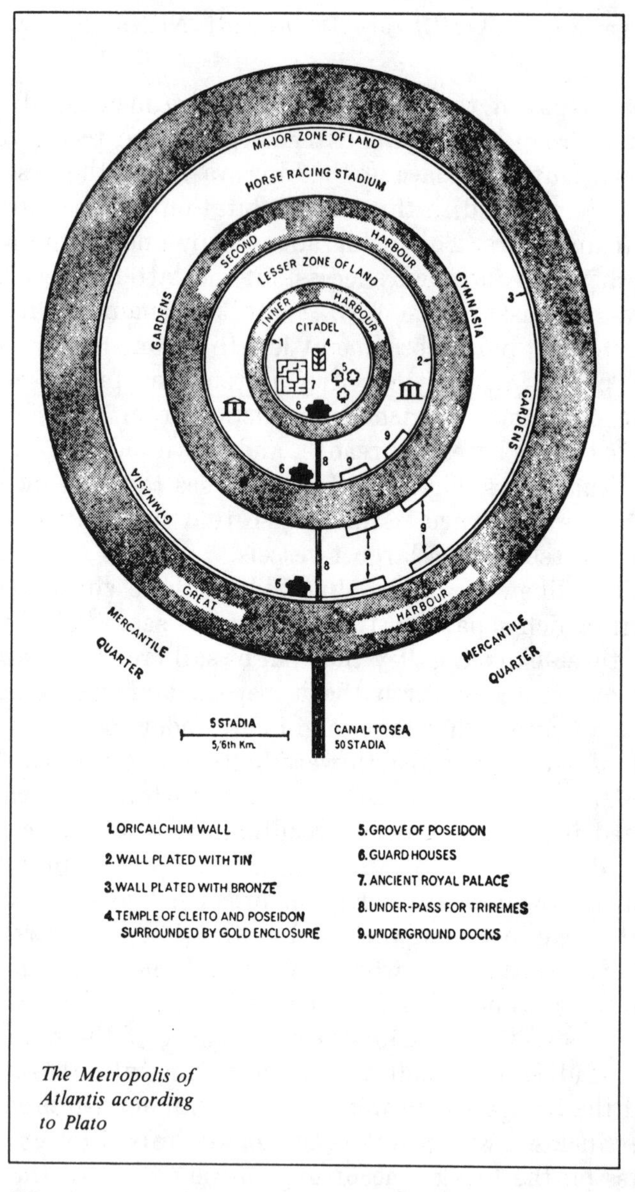

The Metropolis of Atlantis according to Plato. (From *End of Atlantis* by J. V. Luce, Thames and Hudson. Reprinted by permission.)

rock, afforded two convenient docks. With respect to the edifices, some were of a simple structure, and others were raised from stones of different colours; thus by variety pursuing pleasure, which was allied to their nature. They likewise covered the superficies of the wall which inclosed the most outward zone with brass, using it for this purpose as an ointment; but they covered the superficies of that wall which inclosed the interior zone with tin; and lastly, they covered that which inclosed the acropolis with orichalcum, which shines with a fiery splendour.

The royal palace within the acropolis was constructed as follows: In the middle of it there was a temple, difficult of access, sacred to Clites and Neptune, and which was surrounded with an inclosure of gold. In this place assembling in the beginning, they produced the race of ten kings; and from the ten divisions of the whole region here collected every year, they performed seasonable sacrifices to each.

But the temple of Neptune was one stadium in length, and three acres in breadth; and its altitude was commensurate to its length and breadth. There was something, however, barbaric in its form. All the external parts of the temple, except the summit, were covered with silver; for that was covered with gold. With respect to the internal parts, the roof was entirely formed with ivory, variegated with gold, silver, and orichalcum; but as to all the other parts, such as the walls, pillars, and pavement, these were adorned with orichalcum. Golden statues, too, were placed in the temple; and the God himself was represented standing on a chariot, and governing six-winged horses; while, at the same time, through his magnitude, he touched the roof with his head. A hundred Nereids upon dolphins were circularly disposed about him; for at that time this was supposed to be the number of the Nereids. There were likewise many other statues of private persons dedicated within the temple. Round the temple, on the outside, stood golden images of all the women and men that had descended from the ten kings: together with many other statues of kings and private persons, which had been dedicated from the city, and from foreign parts that were in subjection to the Atlantic island. There was an altar, too, which accorded in manner and construction with the other ornaments of the temple; and in like manner, the palace was adapted to the magnitude of the empire, and the decorations of the sacred concerns.

The inhabitants, likewise, used fountains both of hot and cold water, whose streams were copious, and naturally salubrious and pleasant in a wonderful degree. About the fountains, too, edifices were constructed, and trees planted, adapted to these fontal waters. Receptacles of water, likewise, were placed round the fountains, some of which were exposed to the open air, but others were covered, as containing hot baths for the winter season. Of these receptacles, some were appropriated to the royal family, and others, apart from these, to private individuals; and again, some were set apart for women, and others for horses and other animals of the yoke; a proper ornament at the same time being distributed to each.

They likewise brought defluent streams to the grove of Neptune, together with all-various trees of an admirable beauty and height, through the fecundity of the soil: and thence they derived these streams to the exterior circles, by conducting them through channels over the bridges. But in each island of these exterior circles there were many temples to many Gods, together with many gardens, and gymnasia apart from each other—some for men, and others for horses.

But about the middle of the largest of the islands there was a principal hippodrome, which was a stadium in breadth, and the length of which extended round the whole circle, for the purpose of exercising the horses. On all sides of the hippodrome stood the dwellings of the officers of the guards. But the defence of the place was committed to the more faithful soldiers, who dwelt in the smaller circle, and before the acropolis; and the most faithful of all the soldiers were assigned habitations within the acropolis, and round the royal abodes.

The docks, likewise, were full of three-banked galleys, and of such apparatus as is adapted to vessels of this kind. And in this manner the parts about the royal palaces were disposed. But having passed beyond the external ports, which were three in number, a circular wall presented itself to the view, beginning from the sea, and every way distant from the greatest of the circles and the port by an interval of fifty stadia. This wall terminated in the mouth of the trench which was towards the sea. The whole space, too, inclosed by the wall was crowded with houses; and the bay and the greatest harbour were full of ships and merchants that came from all parts. Hence, through the great multitude that were here assembled, there was an all-various clamour and tumult both by day and night. And thus we have nearly related the particulars respecting the city and the antient habitation, as they were then unfolded by the Egyptian priests. In the next place, we shall endeavour to relate what was the nature, and what the arrangement, of the rest of the region.

First, then, every place is said to have been very elevated and abrupt which was situated near the sea; but all the land round the city was a plain, which circularly invested the city, but was itself circularly inclosed by mountains which extended as far as to the sea. This plain too was smooth and equable; and its whole length, from one side to the other, was three thousand stadia; but, according to its middle from the sea upwards, it was two thousand stadia. The whole island, likewise, was situated towards the south, but from its extremities was exposed to the north.

Its mountains were then celebrated as surpassing all that exist at present in multitude, magnitude, and beauty; and contained many villages, whose inhabitants were wealthy. Here, too, there were rivers, lakes, and meadows, which afforded sufficient nutriment for all tame and savage animals; together with woods, various both in multitude and kind, and in abundance adequate to the several purposes to which they are subservient. This plain, therefore, both by nature and the labours of many kings in a long period of time, was replete with fertility. Its figure, too, was that of a square, for the most part straight and long; but on account of the trench which was dug round it, it was deficient in straightness.

The depth, breadth, and length of this trench are incredible, when compared with other labours accomplished by the hands of men: but, at the same time, we must relate what we have heard. Its depth was once acre; and its breadth everywhere a stadium. And as it was dug round the whole plain, its length was consequently ten thousand stadia. [That is, 1,250 miles. This trench, however, was not a more surprising effort of human industry than is the present wall of China.] This trench received the streams falling from the mountains, and which, circularly flowing round the plain towards the city, and being collected from different parts, at length poured themselves from the trench into the sea. Ditches one hundred feet in breadth, being cut in a right line from this part, were again sent through the plain into the trench near the sea: but these were separated from each other by an interval of one hundred stadia. The inhabitants brought wood to the city from the mountains, and other seasonable articles, in twofold vessels, through the trenches; for the trenches intersected each other obliquely, and towards the city. Every year they twice collected the fruits of the earth; in winter using the waters from Jupiter, and in summer bringing the productions of the earth through the streams deduced from the trenches.

With respect to the multitude of men in the plain useful for the purposes of war, it was ordered that a commander in chief should be taken out of each allotment. But the magnitude of each allotted portion of land was ten times ten stadia; and the number of all the allotments was sixty thousand. There is said to have been an infinite number of men from the mountains and the rest of the region; all of them were distributed according to places and villages into these allotments, under their respective leaders. The commander in chief, therefore, of each division was ordered to bring into the field of battle a sixth part of the war-chariots, the whole amount of which was ten thousand, together with two horses and two charioteers: and again, it was decreed that he should bring two horses yoked by the side of each other, but without a seat, together with a man who might descend armed with a small shield, and who after the charioteer might govern the two horses: likewise, that he should bring two heavy-armed soldiers, two slingers, three light-armed soldiers, three hurlers of stones, and three jaculators, together with four sailors, in order to fill up the number of men sufficient for one thousand two hundred ships. And in this manner were the warlike affairs of the royal city disposed. But those of the other nine cities were disposed in a different manner, which it would require a long time to relate.

The particulars respecting the governors were instituted from the beginning as follows: Each of the ten kings possessed authority both over the men and the greater part of the laws in his own division, and in his own city, punishing and putting to death whomsoever he pleased. But the government and communion of these kings with each other were conformable to the mandates given by Neptune; and this was likewise the case with their laws. These mandates were delivered to them by their ancestors inscribed on a pillar of orichalcum, which was erected about the middle of the island, in the temple of Neptune. These kings, therefore, assembled together every fifth, and alternately every sixth year, for the purpose of distributing an equal part both to the even and the odd; and when assembled, they deliberated on the public affairs, inquired if any one had acted improperly, and, if he had, called him to account for his conduct.

But when they were about to sit in judgment on any one, they bound each other by the following compact. As, prior to this judicial process, there were bulls in the temple of Neptune, free from all restraint. They selected ten of these, and vowed to God, they would offer a sacrifice which should be acceptable to him, viz. a victim taken without iron, and hunted with clubs and snares. Hence, whatever bull was caught by them they led to the pillar, and cut its throat on the summit of the column, agreeably to the written mandates. But on the pillar, besides the laws, there was an oath, supplicating mighty imprecations against those that were disobedient. When, therefore, sacrificing according to their laws, they began to burn all the members of the bull, they poured out of a full bowl a quantity of clotted blood for each of them, and gave the rest to the fire; at the same time lustrating the pillar.

After this, drawing out of the bowl in golden cups, and making a libation in the fire, they took an oath that they would judge according to the laws inscribed on the pillar, and would punish any one who prior to this should be found guilty; and likewise that they would never willingly transgress any one of the written mandates. They added that they would neither govern nor be obedient to any one who governed contrary to the prescribed laws of their country. When every one had thus supplicated both for himself and those of his race, after he had drunk, and had dedicated the golden cup to the temple of the God, he withdrew to the supper, and his necessary concerns. But when it was dark, and the fire about the sacrifice was abated, all of them, invested with a most beautiful azure garment, and sitting on the ground near the burnt victims, spent the whole night in extinguishing the fire of the sacrifice, and in judging and being judged, if any person had accused some one of them of having transgressed the laws.

When the judicial process was finished, and day appeared, they wrote the decisions in a golden table, which together with their garments they dedicated as monuments, in the temple of God. There were also many other laws respecting sacred concerns, and such as were peculiar to the several kings; but the greatest were the following: That they should never wage war against each other, and that all of them should give assistance if any person in some one of their cities should endeavour to extirpate the royal race. And as they consulted in common respecting war and other actions, in the same manner as their ancestors, they assigned the empire to the Atlantic family. But they did not permit the king to put to death any of his kindred, unless it seemed fit to more than five out of the ten kings.

Such then being the power, and of such magnitude, at that time, in those places, Divinity transferred it from thence to these parts, as it is reported, on the following occasion. For many generations, the Atlantics, as long as the nature of the God was sufficient for them, were obedient to the laws, and benignantly affected toward a divine nature, to which they were allied. For they possessed true, and in every respect magnificent conceptions; and employed mildness in conjunction with prudence, both in those casual circumstances which are always taking place, and towards each other. Hence, despising every thing except virtue, they considered the concerns of the present life as trifling, and therefore easily endured them; and were of opinion that abundance of riches and other possessions was nothing more than a burthen. Nor were they intoxicated by luxury, nor did they fall into error, in consequence of being blinded by incontinence; but, being sober and vigilant, they acutely perceived that all these things were increased through common friendship, in conjunction with virtue; but that, by eagerly pursuing and honouring them, these external goods themselves were corrupted, and, together with them, virtue and common friendship were destroyed. From reasoning of this kind, and from the continuance of a divine nature, all the particulars which we have previously discussed, were increased among them.

But when that portion of divinity, or divine destiny which they enjoyed vanished from among them in consequence of being frequently mingled with much of a mortal nature, and human manners prevailed; then, being no longer able to bear the events of the present life, they acted in a disgraceful manner. Hence, to those who were capable of seeing, they appeared to be base characters, men who separated things most beautiful from such as are most honourable: but by those who were unable to perceive the true life, which conducts to felicity, they were considered as then in the highest degree worthy and blessed, in consequence of being filled with an unjust desire of possessing, and transcending in power. But Jupiter, the God of Gods, who governs by law, and who is able to perceive every thing of this kind, when he saw that an equitable race was in a miserable condition, and was desirous of punishing them, in order that by acquiring temperance they might possess more elegant manners, excited all the Gods to assemble in their most honourable habitation, whence, being seated as in the middle of the universe, he beholds all such things as participate of generation: and having assembled the Gods, he thus addressed them . . .

APPENDIX B: THE PROBLEM OF RADIOCARBON DATING

In trying to establish a possible chronology at Bimini I encountered various conflicting dates, particularly with the radiocarbon dating established by John Gifford. Gifford took a sample near the surface of the bedrock, earlier identified as biopelsparite, a Pleistocene marine limestone. The uranium-thorium dating method dated the sample at about 15,000 B.P., suggesting that the earliest date for any blocks placed on the present surface would be about 15,000 years ago. Then Gifford found carbon-14 dates of 3200 B.P. for a sample from a block in the seaward lead; a sample from the beachward lead was dated at 2500 B.P. I believe that these dates are too recent.

The carbon-14, or radiocarbon, procedure can be used for dating materials of organic origin as old as 50,000 years. While the actual chemistry is somewhat complex, the method depends essentially on the ratio of C-14, a radioactive isotope, to the inert C-12—a ratio that is identical in all living forms. At death, the C-14 begins to diminish and analysis of the proportion will give an approximate date of the death of the organism. But the older the dates, the more inaccurate they become. An additional inaccuracy, recently discovered, derives from the fact that about 2000 years ago, there was a greater abundance of C-14 (and thus a higher proportion of it in living organisms). This was learned by checking existing C-14 dates against dendrochronology, the absolute system of dating by counting tree rings. The necessary corrections indicate that a C-14 reading of 3500 B.P. is actually closer to 3800 B.P. and a C-14 reading of 5000 B.P. may date from anywhere from 5500–6000 B.P.

Gifford's report of his dates states that samples were taken in such a way as to avoid "bored or incrusted material" that would contaminate the samples (and make them appear younger). He does not comment (as he does elsewhere in reference to another sample) on the degree of recrystallization. The calcium carbonate on which limestone rock is based can take many forms, and these forms change as their environment changes with time. As one crystal structure gives way to another, the apparent date of a sample gets younger. Because we do not know the degree of recrystallization of these two crucial samples, we have no basis for estimating their possible contamination by ground water, which would make them appear younger in laboratory analysis. Furthermore, it is not clear whether corrections based on dendrochronology have been applied to these dates.

The reader can judge the intricacy of the problem from the fact that Gifford made several trips to Bimini even with the support and resources of the University of Miami and the National Geographic Society, and finally produced an interim report in 1971. In 1973, however, he presented this thesis, in which he changed his identification of certain rock types as well as his overall conclusions about the origins of the blocks off Paradise Point. Our own findings, which contradict his, indicate once again that despite good scientific methodology, the past does not readily give up its secrets.

13,000 B.C. This is the date Gifford derived from uranium-thorium dating procedures for the seafloor under the Road at Bimini. At this time the sea level was near its last Ice Age low and the Road would have been about 400 feet above the sea. Could the marine limestone have been formed under these conditions?

4000 B.C. According to Milliman and Emery's Atlantic sea-level figures, the Road would still have been 23 feet above the water. Yet Gifford, (from C-14 dating) says that at this time the main part of the Road formed as beach rock in the intertidal zone (the range of tide on the beach).

2000 B.C. Sea level at Bimini about 8 feet below present level and rising. (C-14 date of peat layer: Newell.)

1200 B.C. Gifford's C-14 date for the seaward of the two shorter rows of megalithic blocks. This is hard to understand, since 800 years earlier the sea appears to have been at least 7 feet above these blocks; in other words, low tide would not likely have exposed them—a condition thought to be a part of the formation of beach rock.

500 B.C. Gifford's C-14 date for the beachward structure of the two shorter rows. Because these blocks are at the same depth, the problem of the 1200 B.C. date is compounded here since the sea is even higher.

APPENDIX C: THE SANTORINI QUESTION

One of the most interesting attempts to reinterpret Plato's legend of Atlantis is the claim by James Mavor and others that Plato was really describing the Minoan culture he and others found on Santorini, an island in the Aegean Sea.

In 1966, John Lear's *Saturday Review* article, "The Volcano That Shaped the Western World," drew my attention to the theory that Plato's Atlantis was located in the eastern end of the Mediterranean Sea, specifically in the Aegean on the island of Thera, or Santorini. In 1939, the Greek archaeologist Dr. Spyridon Marinatos had first suggested that the explosion of Santorini might have ended Minoan civilization. Later, in 1965, this theory was supported by a scholarly paper, "Santorini Tephra," given by Dr. Bruce C. Heezen of the Lamont-Doherty Geological Observatory, and Dragoslav Ninkovich, in England. They cited evidence that about 1400 B.C. the Aegean islands around Santorini had been blanketed with over 10 centimeters of tephra (volcanic ash); such a deposit would have made agriculture impossible. Destruction of the Minoan culture then accelerated the evolution of a late Mycenaean culture on the Greek mainland when Minoan refugees arrived with their art and alphabet. In retrospect, this may have been the real basis of the Golden Age in Greece, and thus the beginning of Western culture itself.

Before the Heezen-Ninkovich paper, another one was delivered in 1960 by Professor Angelos Galanopoulos, director of the seismic laboratory, University of Athens, also relating Santorini to Plato's Atlantis. Galanopoulos contended that Plato had exaggerated both the size and date of Atlantis. To coincide with Santorini's 1400 B.C. explosion, he therefore reduced Plato's date of 9000 years (before Solon's time) by a factor of ten (900 + 590 B.C. [date of Solon's visit] = 1490 B.C.). Galanopoulos also reduced the size of the island continent to fit the Aegean Sea. He kept the metropolis the same size and superimposed a drawing from Plato's description over Santorini, where it seemed to fit. Underwater, the blown-out crater, or caldera, appeared to show a profile of concentric circles similar to Plato's description of the principal city.

Galanopoulos' thesis later attracted the interest of Dr. James W. Mavor, Jr., inventor of the two-man research submarine *Alvin* and researcher at the Woods Hole Oceanographic Institution. As a consequence, during a Mediterranean cruise in 1966, the Woods Hole research vessel *Chain* spent a day sounding Santorini. Mavor was convinced that the preliminary work justified further sounding and coring of the ocean bottom toward Crete.

All of these developments led Lear, *Saturday Review* science editor, to write his very enthusiastic article. Lear said of Mavor's proposed project: "Seldom are the sciences and the humanities equally concerned about a project as deeply as they are in the historical consequences of the great Santorini explosion. Only once or twice in a human lifetime is a dynamic means at hand to remind people that they are evolutionary creatures dependent on the evolution of their environment—the restless planet of a yellow star."

Lear properly reminds us of the crucial need to understand our past existence on this planet in far more dynamic terms. Hence his enthusiasm over Santorini springs from the deep need to recover the racial memories of our ancient past in order that we might see our future more clearly.

The following year, 1967, Mavor's expedition to Thera led to a July 19 *New York Times* headline: "Minoan City, Found After 3,400 Years, Is Linked To Atlantis." Mavor and Dr. Emily Vermeule, professor of art and greek at Wellesley College, found a Minoan city under 30 feet of volcanic ash. They linked the site to Atlantis because of its violent end and its advanced civilization. In his book *Edgar Cayce on Atlantis,* Cayce later called this "the most recent attempt to rationalize the Atlantis legend by changing its location and date." Mavor and Vermeule evidently felt that they had confirmed Galanopoulos' theory. By 1969 Mavor published his *Voyage to Atlantis,* satisfied that he had found it. The thrust of the geological and hydrographic data makes a strong case, *if* we can accept the initial premise that Plato exaggerated the figures of his legend. I am unable to agree, for reasons which the present book should make clear.

APPENDIX D: SACRED GEOMETRY

The examination of any suspected megalithic site involves becoming familiar with the principles of sacred geometry. I found John Michell to be the best source of information, and believe the following statement to be his most significant expression of the essence of sacred geometry:

> The considerations behind the plan and position of the temple were astronomical, geometrical, and numerical, and they were also geological, for the site of the temple was decided by reference to the field of terrestrial magnetism and located where the fusion between the earth current and the forces of cosmic radiation would naturally occur.
>
> Like every art, geometry has a sacred origin, being the visual expression of the canon of proportion. In the practice of sacred geometry the various types of universal motion are represented by shapes and symbols, which in their combination reflect the interaction of creative forces. The synthesis of all is the plan of the cosmic temple.
>
> The cosmic temple was an equilibrator, whose function was to reconcile all the diverse and contradictory aspects of nature. This is also the function of the universe, and the temple was therefore designed as its microcosm. It was the magical control center of all life on earth.

As we have earlier seen, archaeoastronomy applied to European and pre-Columbian sites gives solid evidence of a highly accurate observational astronomy—even in cultures limited to Stone or Bronze Age technologies. This scientific skill was the basis for solar, lunar, and stellar alignments of the temples of such cultures—one essential element of a sacred geometry.

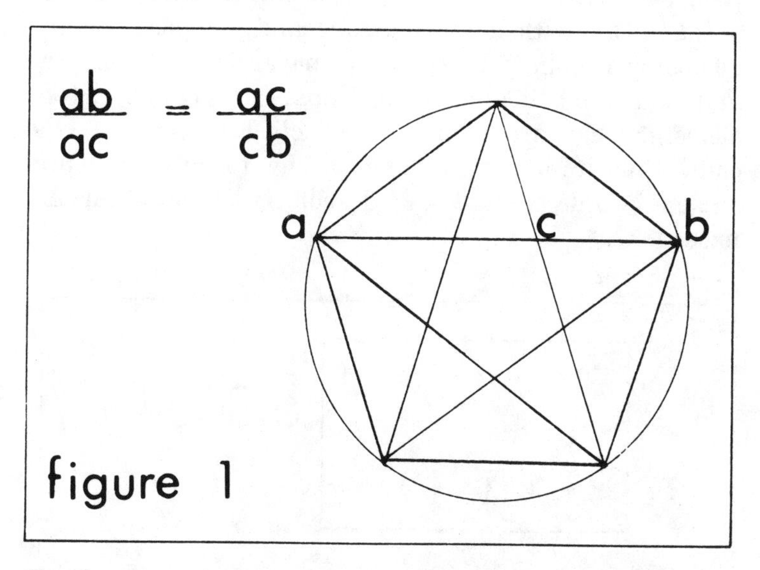

Greek geometric construction of the Golden Section which they called the Golden Mean, the ratio of 1 to 0.618034.

In addition, there are still more subtle aspects of sacred geometry such as the widespread presence of certain geometric shapes and certain numbers, often related, all regarded as sacred in many cultures at least as early as the ancient Egyptians. One writer, Tons Brunés, traces important aspects of sacred geometry back to the Egyptians, from whom he says Pythagoras and later Plato evolved their sacred geometry. In his book, *The Secrets of Ancient Geometry,* he claims that geometry preceded other branches of mathematics in the ancient Egyptian culture and that the circle, square, cross, and triangle—each of which are to be found in the Great Pyramid—were all considered sacred. Brunés says that the ancient geometer began with the circle inscribed within the square and quartered with a cross, then constructed the pentagon, hexagon, octagon, and decagon; all with a straightedge and compass. Sacred geometry is solidly grounded in nature's own geometry and can symbolically relate apparently contradictory elements that are actually reconciled in nature. One could thus infer that, through a now-forgotten science, ancient people created sacred geometry from observing nature.

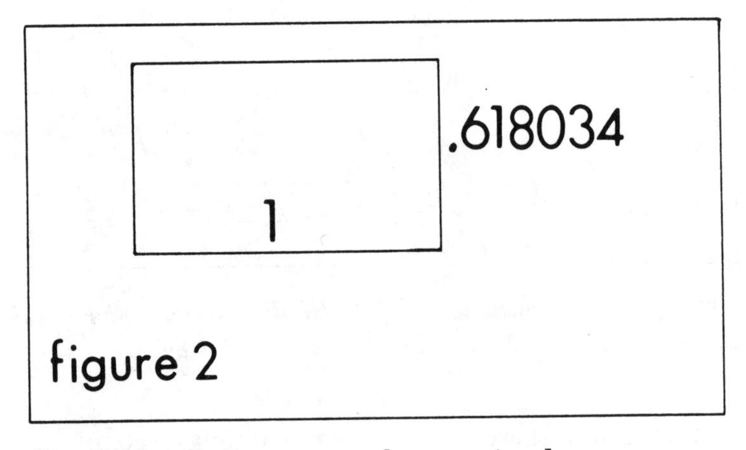

The Golden Section expressed as a rectangle.

It appears that there may be a basis in nature for considering the pentagon the most sacred figure of all. The lines of a five-pointed star drawn within the pentagon (inscribed within the circle) produce the famous Golden Section or the proportion 0.618034 to 1—which the Greeks called the Golden Mean. Apparently lacking the mathematical basis known to the Egyptians, the

Greeks constructed the proportion by geometrical procedures.

Each line of the five-sided star within the pentagon cuts another into two sections so that the smaller section is in proportion to the larger section as the larger section is to the whole line. This proportion, when revived in the Renaissance, came to have an important artistic value. About the same time it began to be recognized as having a widespread distribution throughout nature. When the proportion is expressed in a rectangle with sides equal to 1 and 0.618034 it becomes what Pythagoras and Euclid called the Rectangle of the Divine Section.

In the seventeenth century Jakob Bernouilli noticed that this proportion was the geometric basis for an important curve in nature that often extended into spiral form, which he called the logarithic spiral. In the nineteenth century, the British art critic John Ruskin, seeing this curve on the skyline of the Alps, called it the infinite curve. (As I noted in my doctoral dissertation, Ruskin was unique among his Victorian contemporaries in that he sought to discover an objective psychological basis for aesthetic appreciation.) This curve is found throughout nature: in the cross section of a breaking sea wave, the developmental curve of the chambered nautilus, and the spiral galaxies of the universe.

Beginning with a square, the proportion is arrived at as follows: First the square is divided vertically; then a diagonal AB is constructed that becomes a radius of a curve BC. Taking the square plus its extension DC as 1, then EC equals 0.618034, the Rectangle of the Divine Section. Next a square is constructed in the upper right hand of the rectangle, then in the lower right hand and so on. Connecting the centers of the squares thus constructed gives a series of lines which, when flared out into a smooth curve, establish the logarithic curve so prevalent in nature. Once this geometry and the associated mathematics are understood, it then becomes obvious that nature's beauty (and order) depends upon mathematical principles.

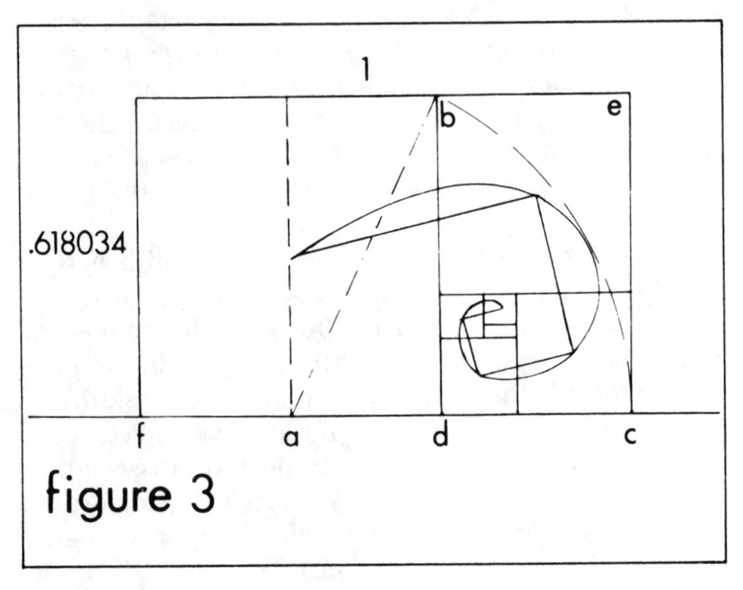

Geometric construction of the logarithic curve from the Golden Section.

As a canon of beauty, the Rectangle of the Divine Section finds expression in everything from the preferred proportion for playing cards to the architectural norm for most rectangular elements such as floor plans, doors, and windows. Whatever the initial basis, the proportion is universally pleasing to our artistic sense. Sacred geometry, then, seems both to please humanity aesthetically and to express an order inherent in nature.

In addition to the Golden Section derived from the pentagon, the numerical element five and pentagonal symmetry seem to be common elements of life on this planet. Philip C. Ritterbush, writing in *The Art of Organic Forms,* tells us that Louis Pasteur's work with polarized light and certain crystals in 1848 demonstrated "that symmetry properties at the molecular level distinguished the living from the non-living." Earlier, Abbé René Just Haüy (1743–1822) had "discovered that the form of crystals was limited by facets whose relation to one another was strictly regular. Thus, for example, a crystal with a pentagonal cross section cannot exist, as a fivefold axis of symmetry cannot be reduced to rational numbers." Pentagonal symmetry, then, is generally found in organic forms, hexagonal in inorganic.

The element of five can be seen in humanity itself: five extremities of the body, five senses, and five races. One of the oldest extant life forms, the shark, has five gills. Many flowers—for example, the rose—exhibit five-sided symmetry.

The number six is closely related to five in various esoteric traditions. Its geometric expansion, the hexagon, is chiefly associated with the inanimate forms of nature, such as the cells of a honeycomb, snowflakes, and crystalline structures. The Giant's Causeway, a natural geological formation located in County Antrim in Northern Ireland, is composed of vertical piles of basaltic rock, each with a six-sided cross section about 18 inches in diameter. Among the ancients the pentagon represented the microcosm, the hexagon the macrocosm; their geometric combination was thus a symbol of the relation between humanity and the universe.

As John Michell demonstrates in *City of Revelation,* the geometric reconciliation of five and six can be accomplished using a figure linked with the Christian mysteries but likely much older, the *vesica piscis* (or vessel of the fish). It is constructed by two equal interpenetrating circles, the center of each located on the circumference of the other. Michell draws upon *Course in the Art of Measurement with Compasses and Ruler,* a work of the German painter Albrecht Dürer, for the

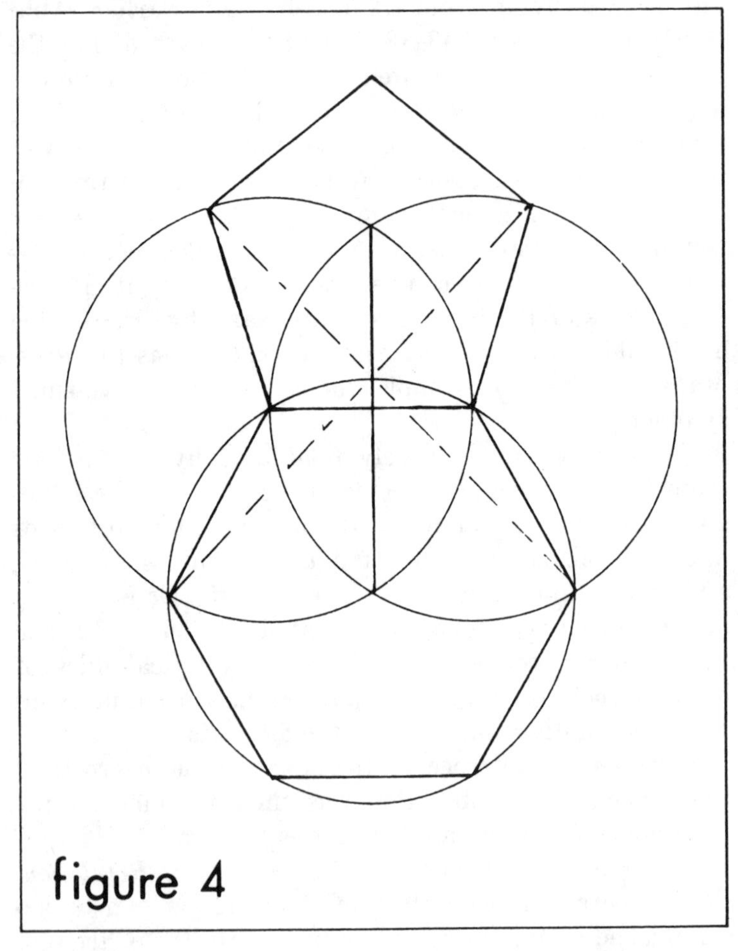

Geometric reconciliation of the numbers five and six.

geometric reconciliation of five and six. In solving the problem, two circles are drawn, their centers at either end of their common radius. Next a line is drawn to connect the intersections of their circumferences. Then a third circle is drawn with its center at the lower intersection of the original circumferences. The two dashed construction lines give the points on the upper circumferences for the pentagon whose apex is the intersection of two more radii. The original radius serves as the first (and common) side of the hexagon and the pentagon. The original radius is also the length of all straight lines in this figure except for the two construction lines.

In Michell's examination of the structure of the Great Pyramid, he found confirmation of the possibility held forth in esoteric traditions of squaring the circle. We must keep in mind that the temple in its construction reconciles not only numerical elements but geometric elements as well.

> The traditions of magic indicate that to construct a square and a circle of equal perimeters, it is necessary first to draw a triangle, and of all triangles the largest and most conspicuous in the world are the four sides of the Great Pyramid, which face the four points of the compass and mark the spot formerly regarded as the center of the earth. The original function of the Pyramid, to promote the union of cosmic and terrestrial forces by which the earth is made fertile, is clearly stated in the symbolism of its geometry, for the Pyramid is above all a monument to the art of squaring the circle.

Following this statement, Michell starts from the Egyptian government survey of 1924, which gave a circumference of 1760 royal cubits (or 3023 feet) and a height of 280 cubits, to explain how the pyramid solves the geometric problem: Begin with a square with a circumference of 1760 cubits that represents the base of the pyramid. Then construct a horizontal line bisecting the vertical sides of the square. Next erect an equilateral triangle whose base angles (from the pyramid) equal 51 degrees and 51 minutes. The altitude of the triangle then becomes the radius of a circle with a circumference approximating the perimeter of the square (2pi $\times$ 280 = 1759.29). Recalling that the pyramid's exact altitude is conjectural because of its missing apex, this is very close.

Even more intriguing is Michell's claim that the solu-

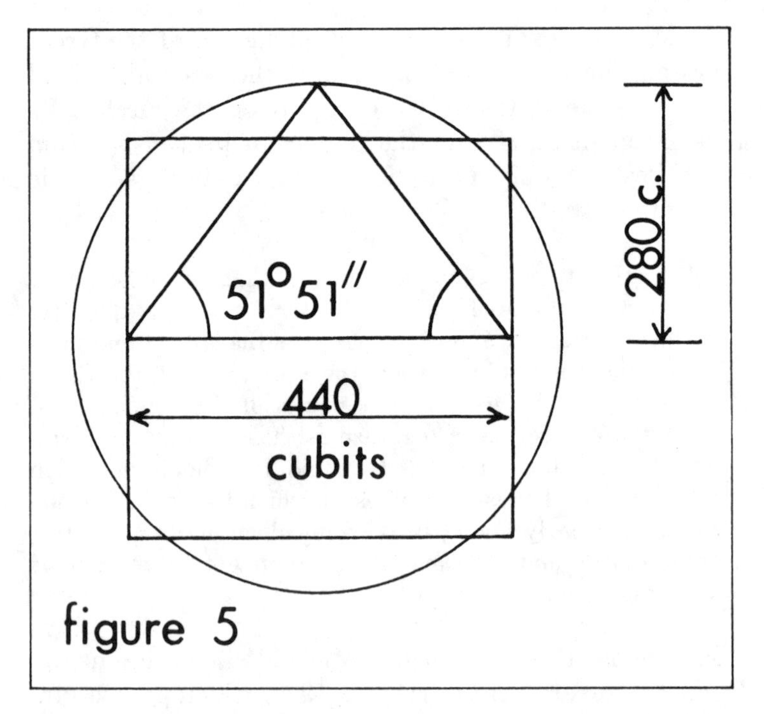

John Michell's geometric solution to the ancient esoteric tradition of squaring the circle.

tion to the problem of squaring the circle is found in nature as well. The actual geometry is not relevant to our immediate purposes, but it seems clear that the relative dimensions of the moon and earth not only provide the solution to squaring the circle but also can be used to generate the sacred right-angle triangle of Pythagoras (a triangle with sides 3, 4, and 5 units long). Once again, we are led to the idea that however the ancients derived it, a sacred geometry appears to be inherent in the natural world.

A sacred geometry's final essential element is the shape of the temple, which functions to resonate with (and amplify) the telluric currents (including the geomagnetic fields, the current induced by running water and so on) as well as cosmic energy, to affect a change of consciousness for those who enter the temple. Again, the Great Pyramid at Giza is the best example. As noted earlier, this structure can even facilitate the out-of-body experience. Apparently the simplest geometry capable of performing this function is the dolmen, a megalithic structure consisting of a tablelike stone which rests upon stone uprights. These are usually thought of as a variation on passage graves found at megalithic sites. In *The Mysteries of Chartres,* Louis Charpentier tells us that the horizontal stone functions as an accumulator of telluric and cosmic energy. Then, because it rests on several (usually three to four) vertical supports, it is under tension and can vibrate like the string of an instrument. "It is," he says, "thus an accumulator and an amplifier." Although Charpentier does not go on to say so, I am quite sure that the structure's interior also functions as a resonant cavity to amplify these subtle energies; it may be analogous to the magnetron that generates microwave energy. Thus these energies are made useful to people, who would not otherwise benefit from them. Once the temple is conceived in all these dimensions, it is obvious that it is not a place for mechanical recitation of words, but a place where actual transformations of consciousness are possible.

APPENDIX E: THE BIMINI MOUNDS

by Raymond E. Leigh, Jr.

A lifetime of interest in the subject of Atlantis led me to plan several trips to Bimini beginning in 1985 but, for various reasons, they never worked out. In March of 1989 I was contacted by Joan Hanley of Boca Raton, Florida, about a Bimini workshop she was putting together. There were twenty people from across the United States who got together on Bimini. In the group was Dr. David Zink, whom I met for the first time. The purpose of the meeting was to investigate the Bimini Road site and fly over the area around Bimini to photograph possible archaeological sites.

The Road site was impressive even when viewed while snorkeling over it; it definitely seemed to be human-made. We dove on the site on April 17, 18, and 19. One of my main reasons for going to Bimini was to get a feel for the island since I was planning an aerial survey of 100 square miles around it. Later I planned to go to the island to provide controls for the aerial photography on which accurate maps could be based to locate future finds. By coincidence, during the week we were there, Dimitri Rebikoff and his crew were doing a photographic survey of the Road site. Joan Hanley arranged for him to fly us over the island and Road site.

The night before our flight, one of our group who was

from Montreal, Paul Heda, told me of some mounds on the island; he had seen them in photographs. This information was exciting to me because of my interest in mounds, particularly the Ohio mounds near where I live in Kentucky. Edgar Cayce, whose work had intensified my interest in Atlantis, had said in one of his readings that the mound builders in the United States were survivors of Atlantis.

On Thursday, April 20, Paul Heda, Bill Donato from California, Mark Furth, a Florida photographer, and I flew with Dimitri Rebikoff, our pilot, over the island. The first site we wanted to see was the Road; immediately after takeoff as we headed for the Road site Bill Donato laughed and said, "There's a fish down there." I caught a glimpse of it out of the plane's window. I asked Dimitri if we could make another pass for a better view. As we flew over it again, I saw a rectangular shape next to the fish. I could not believe that these shapes were in a mangrove swamp! I asked myself, "What is this doing there? I have not read about them in any of the books on Atlantis." We made several other passes overhead. Mark Furth took video shots; the rest of us used still cameras.

We then flew over other parts of the island and saw the straight-line mounds earlier mentioned by Paul Heda, and the Road site. We also looked for a large rectangular shape about 5 miles to the east which I had seen in a 1983 NASA photo, but this bottom pattern was no longer to be seen. Back in Alice Town, Mark ran the videotape repeatedly. We could not believe what we were seeing! That night I met with Nicolas Popov, who was working with Dimitri Rebikoff. Nicolas led exploratory trips in the Bahamas and had a 15-foot rubber raft powered with an outboard motor.

The next day, Friday, we met Nicolas at the Healing

Well and we showed one of the bonefish guides a polaroid photo I had taken of Mark's video. He said he could lead us to the site but was unable to do so. Just then a terrible thunderstorm came across the bank and we had to head back to Alice Town.

I was very disappointed because we were scheduled to leave Bimini the next day, Saturday, at noon. Here it was already 1 P.M. on Friday. As I walked back to my room, I realized that we really had to get out there, whatever the problems. I found Nicolas at his cottage and talked him into taking us around the island to East Bimini. Mark wanted to go to videotape the trip and Patricia Daniels from Virginia wanted to go. We left about 3 P.M. and as soon as we cleared the bar, the winds picked up and we fought 6-foot seas. Several times we almost gave up because of the seas; I was almost seasick, but decided that we had to do it. After another half hour of battling the waves, we got around the northern tip of the island and the seas calmed down. As we motored down the shallow side of the island, Nicolas had to stop frequently to clear the propellor of seaweed.

About 4:30 P.M. we reached the point at which I thought we had to go ashore and work inland. As it was low tide, we had to drag the raft about a half mile. It was not until 5 P.M. that we started chopping our way toward what I thought was the direction of the fish-shaped and rectangular mounds. After twenty-six years of land surveying, I thought I had encountered vegetation as tough as it came, but I was wrong. The Bahamian bush, briars, and mangrove roots are almost impossible to get through. We chopped for about an hour and, at that point, I was ready to turn back. Patti Daniels kept saying, "We can go a little farther can't we?" so we would chop for a few more minutes. The temperature was 93 degrees and the humidity was unbearable because of the morning thunderstorm. Finally, we arrived at a high point from which we could see what looked like a mound about 1000 feet ahead. When I saw the distance yet to go, I was ready to turn back. Nicolas and I shared one small machete as Mark carried the camera. We were chopping mangrove roots 3–4 inches in diameter. I was ready to give it up, but Patti again encouraged us to go ahead.

We finally arrived at the mounds about 6:30 P.M. The mound we first got to was about 10–15 feet high above the floor of the swamp. We climbed to the top of the mound and dug about 3 feet down to try to locate a solid structure, but soon water started entering the hole because of the morning rain. At this point we gave up and headed back to the raft; it was getting very late. Back at the raft we saw that the tide was at the low point and we had to drag it three-quarters of a mile.

Nicolas wanted to go around the southern tip of East Bimini as the seas eastward seemed worse. As it turned out, this was a mistake that forced us to drag the raft for over a mile over sandbars and with small sand sharks all around us. We finally reached deeper water, started the motor and it immediately clogged up with sand. Nicolas radioed for help, and found out how to unclog the motor. Fortunately it worked, because it was another 5 miles to the dock. When we arrived everyone was waiting, afraid that something terrible had happened to us since we were so late. Everyone was very excited about our adventure. We left Bimini the next day and I knew that I would be back soon.

Back home in Elizabethtown, Kentucky, I got my 35-milimeter photos developed and for the next two months analyzed them. Studying the mounds photographed from the air I discovered some strange features on the southeastern portion of the rectangular mound. I was sure that I was seeing some circular objects and some hexagonal objects.

After days of agonizing over these images, my wife, Joyce, insisted that I go back to Bimini and somehow find out what was on this mound. Coordinating this trip with Mark and Christie Furth, I arrived at their home on June 27, arranged for a helicopter flyover of the fish-shaped and rectangular mounds at Bimini on the June 28.

This flight was totally unsuccessful as a photo mission because the low altitude approach agitated the brush so much that all images were blurred. It was a learning experience that cost over 1,000 dollars.

Before this trip I had contacted an aerial survey firm in Miami, Pan American Aerial Surveys. I wanted a professional aerial survey of Bimini Island. The agreement was to complete this survey by June 21, but bad weather delayed the completion until July 28. I had sent the president of the firm, Don Walker, my map with three flight lines drawn on it; he was to fly these flight paths at 1000 feet, 1500 feet, and 2400 feet above sea level. This would give me photo scales of approximately 1 inch = 166 feet, 1 inch = 250 feet, and 1 inch = 400 feet. The sites were to be flown with infrared color photography, which would sense possible underground patterns.

When I received the photographs on July 28, I spent hours analyzing them. At this point, my analysis can only touch the surface of the many strange patterns. There are over a hundred right-angle formations of grasses, waterways, and mounds. There are too many for chance. I have been using aerial photography in my surveying profession for twenty-six years and have never seen right-angle and straight-line formations like

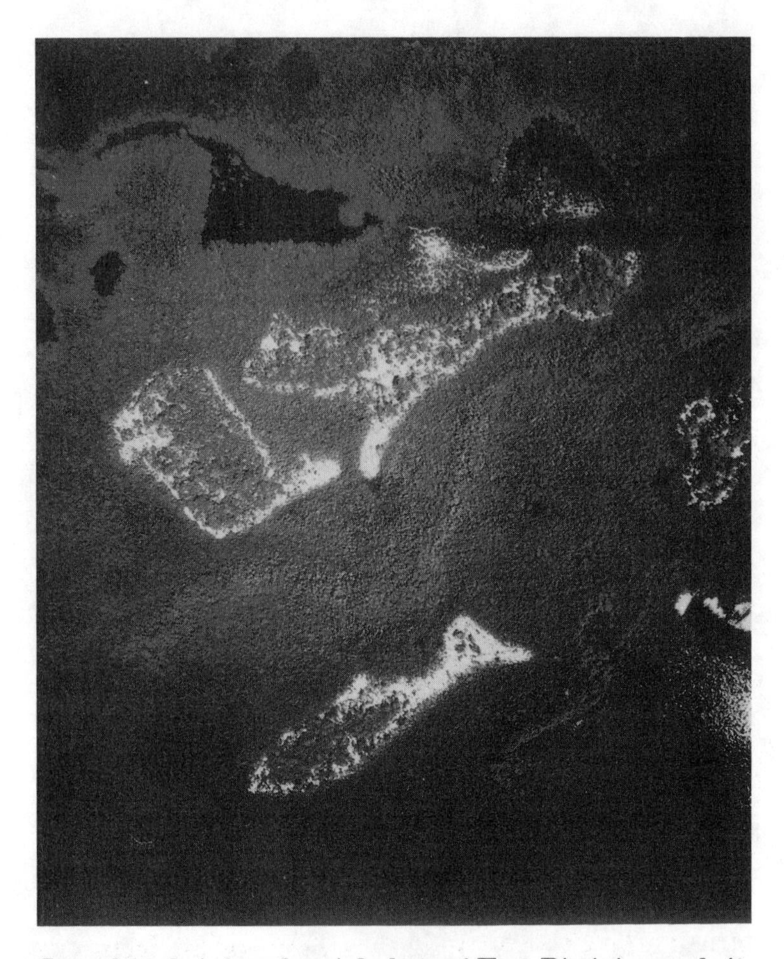

Low altitude infrared aerial photo of East Bimini mound site with the rectangular mound, the cat mound, and the shark mound. (Courtesy Raymond E. Leigh, Jr.)

these in rural areas. The only explanation that I have is that there must be human-made structures under the slime, sand, and mangroves of East Bimini to cause these strange images in the photography.

What these aerial photos show best are the fish mound, which resembles a shark, the rectangular mound, and a newfound mound, which looks like a cat, just northeast of the rectangular mound. The tail of the cat is not a part of the present mound, but the material of which it was made could have been washed away by the numerous hurricanes that have blown over the island. About 600 feet northeast of the shark mound is a straight-line mound approximately 1500 feet in length that has several right-angle formations. In analyzing these images I found some extraordinary mathematical relationships between the shark, rectangular, and cat mounds. Incidentally, a zoologist, Doug Richards, Ph.D., director of research for the ARE, has identified the shark mound as the profile of a lemon shark.

The context for the following analysis is appendix D of the present work. The southeastern edge of the rectangular mound is exactly the same distance from the northeastern edge of the shark mound as the length of the rectangular mound. Taking the divine ratio of Pythagoras and Euclid, 1:0.618, we find the length of the rectangular mound to be 0.618 the length of the shark mound.

If we multiply the length of the shark mound by 1.618, we get the length of the straight-line portion of the end of the cat's tail to the center of the rectangular mound. If we take a point in the center of the rectangular mound and go to the center of the shark mound, then turn northeast (90 degrees left), for the distance between the center of the rectangular mound and the line tangent to the cat's tail, then connect these two points, we will form a rectangle with the divine ratio of 1 to 0.618. If we take the east–west length of the rectangular mound and multiply it by four, we get the distance from the tip of the shark mound to the end of the cat's tail as projected at a right angle from the center of the shark mound. Just northeast of the cat mound there is an additional rectangular formation. All four of these mounds line up with the same general orientation of south 23 degrees west. Finally, the orientation of these mounds differs about 30 degrees from that of the adjacent shoreline.

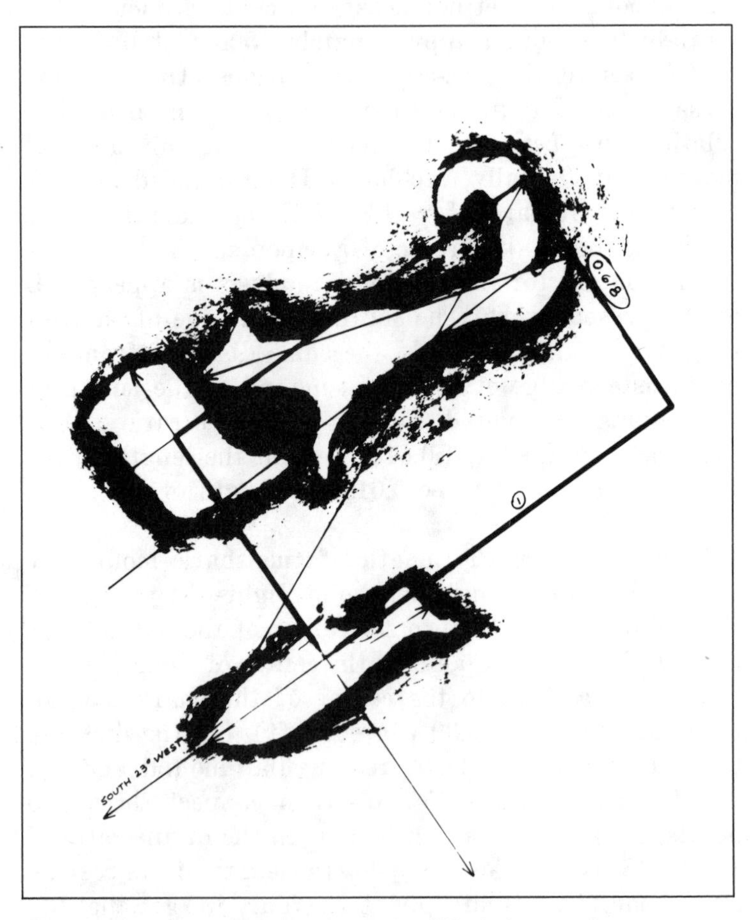

The geometric relationship of the Bimini mounds, 1 to 0.618. (Courtesy Kathleen Lew.)

Because of uncertainty about the height of the aircraft from which the aerial photography was done, it is difficult to give exact dimensions of these mounds. The approximate length of the rectangular mound, however, is 280 feet; its width is 170 feet. The length of the shark mound is 520 feet; its width at the center is 120 feet. The length of the cat mound from nose to tail is 733 feet. And finally, the distance from the mouth of the shark to the cat's tail is 1120 feet.

A 1957 black-and-white aerial photograph of the mound site was also sent to me by aerial photographer Don Walker. Taken at 6000 feet it makes clear that little change has occurred in the last thirty-two years. I think that this is very significant, because if these were natural formations, wind and sea erosion would have changed them.

In Martin Ebon's book on Atlantis, *Atlantis: The New Evidence,* he mentions that Bimini was the site of the Temple to Bastet, the Egyptian cat goddess. It is strange that the aerial photography shows a cat-shaped mound next to a rectangular formation. Was it a temple?

Analysis of a 1968 topographical map of Bimini sent to me by Doug Richards has led me to believe that the entire island of Bimini is a temple complex as claimed by the late Atlantologist, Egerton Sykes. Proof of this will, of course, require extensive excavations by experienced archaeologists. Even before such work, today's technology, including satellite sensing systems, infrared photography, and ground penetrating radar, may help us to find portions of Atlantis under the sand and slime of Bimini.

After being on Bimini for nearly a week, motoring around it in a small raft, flying over it in airplanes and a helicopter, most of it 5–20 feet above sea level, it is difficult to understand why native Indians 850 miles southeast of Bimini in Puerto Rico would have known of a "fountain of youth" on Bimini and would have been able to convince Ponce de Leon of its existence. Without this story, Ponce de Leon would never have reached Florida on April 3, 1513. There must have been a cultural site of extreme importance to support such a legend.

COMMENT

By D.D.Z.

The 1989 Bimini conference planned by Joan Hanley and Vanda Osman at which I shared the discoveries of the ten Poseidia expeditions was a real step forward in cooperation. The work that Raymond Leigh describes above is the most dramatic manifestation of the new cooperative spirit generated by the conference.

As far as the three mounds discovered on East Bimini Island are concerned, the fact that they appeared in infrared aerial photography taken in 1989 essentially as they did in black-and-white aerial photos taken in 1957 argues that they are the work of early humans, not chance sculpturing by natural forces.

Months after Raymond wrote the foregoing essay, he tenaciously continued his painstaking analysis of the aerial photos taken over the island. What next emerged was even more stunning than his recognition of the divine ratio, or what the Greeks called the Golden Section (1 to 0.618) in the geometry of the Bimini mounds.

Suspecting optical distortion in his infrared photos taken at lower altitudes, he returned to the 1957 photo taken at 6000 feet and established its scale from a current Bahamian government topographical map: 1 inch = 1000 feet. He then consulted the latest edition of

John Michell's book, *The New View over Atlantis* (1983).

The result of this analysis was an astonishing conclusion: within 1 percent of the currently accepted measurement of the Egyptian royal cubit, he found that the Bimini mounds not only exhibited a mathematical relationship of 1 to 0.618 but that the unit of measure that best fit the mounds was the same unit of measure as used at Teotihuacán, north of Mexico City, Stonehenge in England, the ancient Temple of Jerusalem *and* the Great Pyramid at Giza!

This unit is one half of what John Michell calls the "sacred rod" of 3.4757485 feet, which equals one part in 6 million of the earth's polar radius *and* the width of the lintel stones at Stonehenge. One half of this measure is 1.73787 feet, the unit that Raymond found to fit the scaled mound dimensions; 1.73787 feet is within 1 percent of the royal cubit of Memphis, 1.7196431. The unit Raymond used produced dimensions for the rectangular mound of 100 by 161.8, a distance between the rectangular mound and the shark of 200 units, and a length of 300 units for the shark mound.

Because of the precision of this analysis, I am convinced that it will stand scrutiny by other professionals. Of course, the analysis of aerial photography must be verified by archaeological investigation of the mounds themselves. The critics of the underwater site off North Bimini will not, however, find an easy disclaimer this time. Clearly, Bimini is an archaeological site once occupied by a yet-to-be discovered New World culture.

At this point, the only other clues to possible cultural connections are a 1000-year-old (or older) hardwood panther artifact from Key Marco in the Florida Keys, and speculations about early Egyptian voyagers who might have brought worship of the Cat Goddess of the Nile, Bastet, to Bimini.

Whatever the cultural links, the stunning facts are the evidence for the ancient proportion of 1:0.618 in the geometric relationships found in the mounds *and* the presence of an ancient unit of measure found on both sides of the Atlantic Ocean.

APPENDIX F: THE LABYRINTH AS A SACRED SITE

Whether found in architectural form or drawn as a maze (as in the case of the thirteenth-century maze found on the floor of Chartres Cathedral), the labyrinth symbolizes human movement through a perplexing, intricate passageway to the revelation of mystery. In *The Way of the Sacred,* Francis Huxley, commenting on the symbolic meaning of the mosaic mazes on old Christian church floors as "the pilgrimage," notes that "the tradition is incalculably old." He goes on to say that "ancient mazes marked out in turf or stones are often called after Troy, itself a word apparently meaning 'to turn'. These Troy mazes are connected with dances at places held to be entrances to the other world. This is why he who knows how to follow or make the diagram has his passport to the other world and resides in the god—or, rather, because the maze honors woman and the belly, the goddess."

The maze has been found even in a contemporary primitive culture in the western Pacific. In *Stone Men of Malekula,* anthropologist John Layard tells of a megalithic culture in the New Hebrides whose sacred site includes altars, monoliths, and dolmens remarkably similar to those in European sites. Their most important ceremony, the Journey of the Dead, involves the artistic motif of the labyrinth drawn as a geometric maze. Layard sees the Malekulan ceremony as parallel to the Egyptian funerary monuments in which "tomb, labyrinth and portrait statue of the dead" are juxtaposed. He adds that "the motivation for the Journey of the Dead is to be sought not in the fact of death itself, but in the desire for the renewal of life through contact with the dead ancestors who are already leading a life beyond the grave." The most exotic labyrinth of the past is the Cretan one created by the legendary architect Daedalus to house the Minotaur, the monster later slain by Theseus. Unless the legend referred to the complex floor plan of the Royal Palace of Knossos, this labyrinth has not yet been found, although its diagram appears on coins from Knossos as late as the fourth century B.C. The Roman scholar Pliny wrote of this labyrinth and three others that modern archaeology has more or less accounted for.

The best-known historically was the Egyptian labyrinth east of Lake Moeris, which both Herodotus and Strabo visited. The twelfth dynasty pharaoh Amenemhet III is credited with the construction (about 1800 B.C.). Flinders Petrie located its 1000- by 800-foot foundations in 1888; (Egyptologists say that the word labyrinth is Egyptian in its origins, meaning "the temple at the entrance of the lake.") This particular labyrinth must have been very impressive. A single wall surrounded 12 courts and 3000 chambers, half of which were underground. Herodotus, who was not allowed to visit those underground chambers, was told that they contained the tombs of the kings who built it and the sacred crocodiles.

Two lesser-known labyrinths described by Pliny are located on the Greek island of Lemnos and in Italy. The former, with 150 columns, is thought to be modeled after the Egyptian site. The latter, Clusium, said to have been built by the Etruscan king Lars Porsena for his tomb, has been identified with a mound named Poggio Gajella near Chiusi.

An important meaning of both the maze and the labyrinthine structure has been identified by J. E. Cirlot in his *Dictionary of Symbols.* It is "the loss of the spirit in the process of creation—that is the 'fall' in the Neoplatonic sense—and the consequent need to seek the way out through the 'Center' back to the spirit." The Bimini psychic readings suggest that this was the exact ceremonial function of the Bimini site. Cirlot, drawing upon the work of Mircea Eliade, the renowned French authority on myth and symbols, adds that "the essential mission of the maze was to defend the 'Center'—that it was, in fact, an initiation into sanctity, immortality and absolute reality (in the Platonic sense) and, as such, equivalent to other 'trials' such as the fight with the dragon (the more subtle meaning, then, of the minotaur). At the same time the labyrinth may be interpreted as an apprenticeship for the neophyte who would learn to distinguish the proper path leading to the Land of the Dead."

In one of her readings during the 1975 expedition, Carol offered another possible explanation of the labyrinth's symbolism: "The winding and the labyrinth is a representation of the planet revolving around a central core, a representation of the winding from the inner part of this galaxy out and then moving back in." This brought to mind the spiral motif, found in many early cultures but particularly important in the petroglyphs of Irish megalithic sites. So the spiral is perhaps another expression of the labyrinth, a place of initiation. Perhaps the fret (the stairs to the stars), a prominent motif in Tiahuanaco pottery, is another version of the initiation or raising of consciousness from one level to another. The labyrinth also seems to be a key to cosmological understanding.

None of the more subtle history of the labyrinth was known to me at the time of the expedition, nor did Carol seem to have knowledge of it. Yet her subsequent reading produced an account of the sacred functions of the site entirely consistent with these traditional symbolic and spiritual functions of the labyrinth. Perhaps we were fortunate enough to have discovered the original source of this important idea. Immediately after the expedition, Joan did a reading on the sacred geometry of the site that even alluded to crocodiles on Bimini. This was some six months before the labyrinth and crocodile had been connected historically in my research.

APPENDIX G: A FOUNTAIN OF YOUTH BREAKTHROUGH

by Richard Wingate

In 1612, an Indian legend was translated that described "An islande about three hundred and twenty five leagues from Hispaniola . . . [Haiti and the Dominican Republic] called Boiuca or Agnaneo, in which is a sprynge of running water of such marvelous vertue that the water thereof being dronke, perhappes with some diette, maketh owlde men yonge again." The island of Boiuca or Agnaneo was also known as Bemene.

Juan Ponce de Leon, who had just been ousted from the lucrative governorship of Puerto Rico, managed to obtain a grant from the King of Spain to colonize and exploit Florida and the Bahamas. He believed the legend enough to sacrifice a vast fortune in Spanish gold in a futile search for the island of Bemene and its legendary Fountain of Youth. Fruitlessly he visited almost every large or important island in the Bahamas, but somehow managed to miss the two tiny islands of Bimini.

In 1926, Edgar Cayce, while in a deep trance, predicted that the remains of an ancient temple would be found underwater near these Bahamian islands. He described Bimini as having been the "highest portion left above the waves of a once great continent . . . Atlantis." Cayce also predicted that one of the biggest health spas in the world could be built on Bimini to tap the healing waters that rejuvenated the people of ancient Atlantis and helped them live to the satisfying age of 200 years or more. Cayce said that wells could be drilled from which would flow marvelous healing waters that would regenerate the sick and give comfort to the aging.

Melaney Freeman, one of the first licensed woman pilots in the United States, subsequently went to Bimini on Cayce's advice and wired back that she had found a freshwater well, walled around with stones of an odd composition and carved with strange hieroglyphic symbols. Unfortunately, she dropped out of sight, and those who have gone to Bimini in search of this marvelous fountain have failed to find it. A subsequent hurricane must have buried it in mud, as it has not been relocated.

The modern part of our story begins in September 1970, at Elsinore, California. Because most healing water found throughout the world contains valuable minerals, I suspected that Edgar Cayce's Bimini well might be rich in minerals. Why not try to locate the well with the help of a dowser? I was led to a Californian named Verne Cameron, who reputedly had more experience locating hot and cold mineral water than anyone else. Verne also had exhibited an uncanny talent for locating such diverse materials as kaolin pottery clay for a ceramics manufacturer in Oaxaca, Mexico; gold mines in the American southwest, oil wells for professional oil promoters, and radium deposits for mining companies. He even dowsed a map of New Zealand and located 85 percent of the oil bearing strata now being worked there today. Most amazing of all, Verne did all his New Zealand locating from a map in Elsinore, California. Documentary evidence for the New Zealand dowsing appears in his book *Aquavideo,* published by El Cariso Press.

Verne worked with an aurameter, a special divining rod he invented, and "Peter," his spirit guide. Using a map of Bimini, Peter guided Verne's hands to a spot inside the harbor on the north island where the water would be found to be curative, regenerative, and healthful. This water could be tapped within 20 feet below the bottom and would get hotter and more pure as we drilled down. The ideal temperature for bathing would be 98 degrees Fahrenheit, which we could expect to hit at only 200 feet. Verne marked a 32-foot square on the map and seemed to think it was an ancient bathing pool, perhaps used by the people of Atlantis to rejuvenate themselves.

Ten days later, Verne and I flew to South Bimini, where we took the small ferryboat that plies the channel between the north and south islands. As we were leaving the south island, Verne's aurameter came alive and pointed to the middle of the wide, shallow harbor. The ferry moved past the small mangrove cay, and as we headed for the dock, the aurameter turned completely around in Verne's hands and continued to point toward the small mangrove cay that now lay behind us.

The next day, we set out on our own to investigate the area. As we approached the cay, the aurameter in Verne's hand began to dance and wave back and forth. Verne verified the map readings he had done in Elsinore by walking over the mangrove cay, and to our delight they indicated a 30-foot-square bathing pool resting on a deep fissure. Peter said a tiny amount of the water was still seeping up through the 12 feet of sand covering the well, but that the bay tides were greatly diluting it.

We flew back to Miami to plan the drilling and pumping operation. Verne questioned Peter about the water, and was told that this was indeed the long lost Fountain of Youth, that the water was hot, curative, rejuvenating, and mineral-rich, but that neither the minerals nor the heat were the chief curative ingredient.

If not the minerals and the heat, what? Then I remembered that Pierre Curie and A. Laborde had discovered radioactive gases in many of Europe's healing springs. One, only 20 miles from the healing waters of Lourdes, is still giving off radioactive gases as it probably has for thousands of years. When Verne quietly asked Peter if our Bimini well was radioactive, the aurameter indicated yes! Of course there are reams of medical evidence to verify radioactivity's healing influence on the circulation. Victor B. Ott, in his chapter on cardiovascular disorders in *Medical Hydrology,* notes that alpha radiation appears to have an ionizing effect that causes rejuvenation of human cells.

Later, when water samples from different parts of the bay were analyzed, we found that the bay water did, indeed, contain small amounts of alpha radiation. The samples taken with weighted sample bottles directly over the suspected 30-foot-square well yielded significantly higher amounts of radiation. Verne had indeed picked a "hot spot." The first proof of the water's affect came months later when Georgina Thompson, a well-known model, told me that while helping with the project and diving in the water over the hot spot, her skin tone had improved, and in spite of the glaring sun and salt water she had been exposed to, small wrinkles had disappeared from around her eyes. Her hair had softened and become more manageable. She was so convinced of the change for the better that I began looking more closely at my own skin. Yes, it did seem to be getting smoother.

I was later told that the water would have a favorable effect on the glands of the body, and hence, stimulate beneficial hormone production. It should also rejuvenate the pituitaries and the adrenal complex, produce profound changes for the better and increase the aged's pep and sense of well-being. Was there any danger? I did wonder, as radiation in the wrong quantities and wrongly used can cause frightening damage. But I learned that this water would be safe if taken in gradually increasing doses two weeks on and two weeks off for a total of about eight weeks.

Peter indicated that our water was emitting radon gas and a host of other radioactive materials. Verne flew back to Elsinore, and I settled down to earning a living and planning to drill for the water. Unfortunately, Verne passed away a short time later, and we never got a chance to drill on Bimini together.

CHRONOLOGY OF THE POSEIDIA EXPEDITIONS

(Led by David D. Zink in the Bahamas)

1974 Six weeks of reconnaissance at Bimini sailing *Makai II* out of Galveston. In Miami Dr. J. Manson Valentine gives Zink a full briefing including slides taken from the air. An initial tape-and-compass survey is done on the Road. Fathometer profile from seaward to beachward leads of Road indicate a roughly horizontal seafloor. Beachrock is usually formed on a slope.

1975 Seventy days of field work at Bimini based on *Makai II* and *Gypsy.* Tape-and-compass survey of underwater Road site. East Site probed down to 9 feet with an airlift; no stonework found. Two artifacts found: a tongue-and-groove building block and a stylized marble head. Peter Tompkins films the expedition. His documentary, filmed during one week of a seventy-day project, was a portion of the film that sold Alan Landsburg's "In Search Of" series to NBC. John Steele, archaeologist; John Parks, geologist. During Tompkins' filming, Dimitri Rebikoff deploys his *Pegasus* and Count Pino Turolla dives on the site.

1976 January: Karen G., Frank Auman, Bill Beidler flown in by Frank Auman to attempt a psychic search for "lighted column"; attempt is not successful.

March: Ultraviolet photography from Sunshine Inn on bearings suggested by Karen G. No significant images.

April: Zink named Explorer of the Year by International Explorers Society of Coral Gables, Florida, for fieldwork at Bimini.

July: Zink invited as on-camera consultant by Cousteau Society to film on Bimini site. Result is "Calypso's Search for Atlantis" filmed by Philippe Cousteau from Cousteau Society's PBY. Major discovery: that the Andros "temple" is really a sponge kraal, or holding pen, built in the 1930s.

August: *Sea Fiddle,* Capt. Josh Sherman, USN (Ret.), supervises survey of the Bimini Road. Survey begins with sun line established on the beach with a transit and buoys over key points of 1975 underwater tape-and-compass survey which are then plotted. Doug Richards assists in the survey.

September: Zink elected a Fellow (nonresident) by The Explorers Club, New York City, in recognition of Bimini fieldwork.

1977 June: *Martech,* Dr. Raymond McAllister supervises coring of blocks at Road site. Marble head hoisted and delivered to Bimini commissioner.

July: *Shepahoy,* Joseph Libbey, divemaster from Smithsonian Institution. Continue search for the lighted column off South Bimini.

1978 January: *Margeo IV,* Dr. Harold Edgerton. Side-scan sonar survey of Road and west of North Bimini from Paradise Point to harbor entrance. Numerous rectilinear targets recorded on strip chart off North Bimini Island, especially from Crossing Rocks south to Entrance Point.

February 17: Nuclear activation analysis of twelve cores taken from the Road in 1977 by Alan P. Curtner at Virginia Polytechnic Institute and State University. Significant differences observed between megalithic blocks and seafloor in terms of trace elements.

Poseidia project described in *In The Spirit of Enterprise: From the Rolex Awards,* foreword by Gerard Piel, published by *Scientific American* (San Francisco, W.H. Freeman).

The Stones of Atlantis by David D. Zink (Englewood Cliffs, N.J.: Prentice Hall) published, an account of the first seven expeditions. Editor: Claire Gerus.

1979 April 9: Lecture, "Underwater Enigma at Bimini Islands, Bahamas," given at The Explorers Club, New York City. Zink assisted by Lt. Paul Evancoe, who describes the forthcoming Poseidia project.

June: *Lark,* official cooperation with U.S. Navy and Defense Mapping Agency. Metal returns from inner curve of J, radioactivity and geothermal activity off South Bimini. Piece of worked stone found in cave underwater at Crossing Rocks. Terry Mahlman, archaeologist.

The Ancient Stones Speak by David D. Zink (New York: E. P. Dutton) published. Two additional expeditions included.

1980 With *SEAS;* Zink's group unable to reach their objective due to a cold front from the north and a tropical storm WSW of Key West. The group led by Tal Lindstrom continues to explore Bimini area. Middle Shoals objective abandoned on November 10, 1980.

1982 With coauthor Terry Mahlman, Zink gives paper on the Poseidia expeditions to Conference on Underwater Archaeology at University of Pennsylvania.

1988 About 80 percent of Zink's correspondence and maps become part of the World Exploration collection at the University of Wyoming's American Heritage Center in Laramie. Projects include Bimini and Honduras (La Ciudad Blanca).

BIBLIOGRAPHY

BOOKS

Aveni, Anthony F., ed. *Archaeoastronomy in Pre-Columbian America.* Austin, Texas: University of Texas Press, 1975.

Berlitz, Charles. *Atlantis: The Eighth Continent.* New York: Ballantine, 1985.

———. *Mysteries from Forgotten Worlds.* New York: Doubleday and Co., 1972.

Boland, Charles M. *They All Discovered America.* New York: Doubleday and Co., 1961.

Braghine, Colonel Alexander Pavlovitch. *The Shadow of Atlantis.* New York: E. P. Dutton, 1940.

Brinton, Daniel G., M.D. *Myths of the New World.* Introduction by Paul M. Allen. Blauvelt, New York: Multimedia Publishing Corporation, 1976.

Bronowski, Jacob. *The Ascent of Man.* Boston, Toronto: Little, Brown and Co., 1974.

Calder, Nigel. *The Restless Earth: A Report on the New Geology.* New York: Viking Press, 1972.

Carnac, Pierre. *L'histoire commence à Bimini.* Paris: Laffont, 1973.

Cayce, Edgar Evans. *Edgar Cayce on Atlantis.* New York: Paperback Library, 1968.

Cayce, Edgar Evans, Gail Cayce Schwartzer, and Douglas G. Richards. *Mysteries of Atlantis Revisited.* San Francisco: Harper & Row, 1988.

Ceram, C. W. *Gods, Graves and Scholars.* 2d rev. ed. E. G. Garside and Sophie Wilkins, trans. New York: Alfred A. Knopf, 1970.

Charpentier, Louis. *The Mysteries of Chartres Cathedral.* Ronald Fraser, trans. New York: Avon Books, 1975.

Charroux, Robert. *Forgotten Worlds.* Lowell Blair, trans. New York: Popular Library, 1973.

Cirlot, J. E. *A Dictionary of Symbols.* Jack Sage, trans. New York: Philosophical Library, 1962.

Craton, Michael. *A History of the Bahamas.* 2d rev. ed. London: Collins, 1968.

David, A. Rosalie. *The Egyptian Kingdoms.* Oxford: Elsevier-Phaidon, 1975.

de Camp, L. Sprague. *Lost Continents: The Atlantis Theme in History, Science, and Literature.* 1954. Reprint. New York: Dover Publications, 1970.

de Santillana, Giorgio, and Hertha von Deschend. *Hamlet's Mill: An Essay on Myth and the Frame of Time.* Boston: Gambit Inc., 1969.

Donnelly, Ignatius. *Atlantis: The Antediluvian World.* Egerton Sykes, ed. New York: Gramercy Publishing Co. with Harper & Row, 1949.

Eiseley, Loren. *The Invisible Pyramid.* New York: Charles Scribner's Sons, 1970.

———. *The Night Country.* New York: Charles Scribner's Sons, 1971.

Eisler, Riane. *The Chalice and the Blade: Our History, Our Future.* San Francisco: Harper & Row, 1987.

Fairbridge, R. W. "Eustatic Changes in Sea Level," in *Physics and Chemistry of the Earth,* vol. 4. New York: Pergamon Press, 1961.

Fawcett, P. H. *Lost Trails, Lost Cities.* New York: Funk and Wagnalls, 1953.

Ferro, Robert, and Michael Grumely. *Atlantis: The Autobiography of a Search.* New York: Doubleday and Co., 1970.

Fix, William R. *Pyramid Odyssey.* Urbanna, Virginia: Mercury Media, 1984.

Garvin, Richard. *The Crystal Skull.* New York: Doubleday and Co., 1973.

Gimbutas, Marija. *The Goddesses and Gods of Old Europe, 7000–3500 B.C.* Berkeley and Los Angeles: University of California Press, 1982.

Goodman, Jeffrey D. *Psychic Archaeology.* New York: G. P. Putnam's Sons, 1977.

Gordon, Cyrus H. *Before Columbus: Links Between the Old World and Ancient America.* New York: Crown Publishers, 1971.

Graves, Robert. *The Greek Myths.* 2 vols. New York: Braziller, 1959.

———. *The White Goddess.* New York: Farrar, Straus and Cudahy, 1948.

Gribbin, John, ed. *Climatic Change.* Cambridge and New York: Cambridge University Press, 1978.

———. *The Hole in the Sky: Man's Threat to the Ozone Layer.* New York: Bantam New Age, 1988.

Hadingham, Evan. *Secrets of the Ice Ages: The World of the Cave Artists.* New York: Walker and Co., 1979.

Hapgood, Charles H. *Maps of the Ancient Sea Kings: Evidence of Advanced Civilization in the Ice Age.* Philadelphia: Chilton Books Co., 1966.

———. *The Path of the Pole.* Rev. ed. of *Earth's Shifting Crust,* 1958. Philadelphia: Chilton Books Co., 1970.

Hawkes, Jacquetta. *Atlas of Early Man.* New York: St. Martin's Press, 1976.

Hawkins, Gerald S. *Beyond Stonehenge.* New York: Harper & Row, 1973.

———. *Stonehenge Decoded.* New York: Doubleday and Co., 1965.

Hitching, Francis. *The World Atlas of Mysteries.* London and Sydney: Pan Books, 1976.

Hoyle, Fred. *From Stonehenge to Modern Cosmology.* San Francisco: W.H. Freeman, 1972.

Jacobi, Jolande. *The Way of Individuation.* R. F. C. Hull, trans. New York: Harcourt, Brace and World, 1967.

Jobes, Gertrude and James. *Outer Space: Myths, Name Meanings, Calendars; From the Emergence of History to the Present Day.* New York and London: Scarecrow Press, 1964.

Kuhn, Thomas S. *The Structure of Scientific Revolutions.* Vol. 2, no. 2 of *International Encyclopedia of Unified Science.* Chicago: University of Chicago Press, 1962.

Levi-Strauss, Claude. *The Raw and the Cooked: Introduction to a Science of Mythology.* John and Doreen Weightman, trans. New York: Harper & Row, 1969.

Lovelock, James. *The Ages of Gaia.* New York: W. W. Norton, 1988.

Marshack, Alexander. *The Roots of Civilization.* New York: McGraw-Hill, 1972.

Masters, R. E. L., and Jean Houston. *The Varieties of Psychedelic Experience.* New York: Holt, Rinehart and Winston, 1966.

Mavor, James W., Jr. *Voyage to Atlantis.* New York: G. P. Putnam's Sons, 1969.

Merz, Blanche. *Points of Cosmic Energy.* Saffron Walden, England: C. W. Daniel Co., Ltd., 1987.

Michell, John. *City of Revelation.* New York: Ballantine Books, 1972.

———. *The View Over Atlantis.* New York: Ballantine Books, 1969.

Ostrander, Sheila, and Lynn Schroeder. *Psychic Discoveries Behind the Iron Curtain.* Englewood Cliffs, New Jersey: Prentice-Hall, 1970.

Ponte, Lowell. *The Cooling.* Englewood Cliffs, New Jersey: Prentice-Hall, 1976.

Renfrew, Colin. *Before Civilization: The Radiocarbon Revolution and Prehistoric Europe.* New York: Alfred A. Knopf, 1973.

Ritterbush, Philip C. *The Art of Organic Form.* Washington, D. C.: Smithsonian Institution Press, 1968.

Roberts, Anthony. *Atlantean Traditions in Ancient Britain.* Llanfyndd, Carmarthen, Wales: Unicorn Bookshop, 1974.

Solomon, Ralph S. *Shanidar.* New York: Alfred A. Knopf, 1971.

Spence, Lewis. *Atlantis in America.* Detroit: Springing Tree Press, 1972.

———. *Atlantis Discovered.* Reprint of *The Problem of Atlantis.* New York: Causeway Books, 1974.

Stearn, Jess. *Edgar Cayce: The Sleeping Prophet.* Garden City, New York: Doubleday and Co., 1967.

Steiger, Brad. *Atlantis Rising.* New York: Dell, 1973.

Stevenson, Ian, M.D. *Twenty Cases Suggestive of Reincarnation.* New York: American Society for Psychical Research, 1966.

Steward, Julian H., ed. *The Circum-Caribbean Tribes.* Vol. 4 of *Handbook of South American Indians.* New York: Cooper Square Publishers, 1963.

Sykes, Egerton. *Bibliography of Classical References to Atlantis.* Rome, 1945. Sykes was editor of the journal *Atlantis.*

Teilhard de Chardin, Pierre. *The Phenomenon of Man.* New York: Harper & Row, 1965.

Thom, Alexander. *Megalithic Lunar Observatories.* New York: Oxford University Press, 1971.

———. *Megalithic Sites in Britain.* New York: Oxford University Press, 1967.

Tompkins, Peter. *Mysteries of the Mexican Pyramids.* New York: Harper & Row, 1976.

———. *Secrets of the Great Pyramid.* New York: Harper & Row, 1971.

Ushakov, S. A. *Physics of the Earth.* Vol. 1 of *The Structure and Development of the Earth.* A. P. Kaplan, USSR Academy of Sciences, series ed. Boston: G. K. Hall and Co., 1977.

Velikovsky, Immanuel. *Worlds in Collision.* New York: Macmillan, 1950.

Wuthenau, Alexander von. *Unexpected Faces in Ancient America (1500 B.C.–A.D. 1500).* New York: Crown, 1975.

Ywahoo, Dhyani. *Voices of Our Ancestors: Cherokee Teachings from the Wisdom.* Barbara DuBois, ed. Boston and London: Shambala, 1987.

Zhirov, N. F. *Atlantis.* David Skvirsky, trans. Moscow: Progress Publishers, 1970.

Zink, David D. *The Ancient Stones Speak.* New York: E. P. Dutton, 1979.

———. *Leslie Stephen.* New York: Twayne, 1972.

———. *The Stones of Atlantis.* 1st ed. Englewood Cliffs, New Jersey: Prentice-Hall, 1978.

Zink, Joan, and David D. *You Are the Mystery.* Lakemont, Georgia: CSA Press, 1976.

ARTICLES

Abrahamsen, N. "Do the Pyramids Show Continental Drift?" *Science* 179 (March 1973): 892–93.

Chapin, Brady. "A Study of Adjacent Samples Taken from a Beachrock Formation off Paradise Point, Bimini, Bahamas," (1976). Independent study for Dr. Herbert Tischler, Earth Sciences Department, University of New Hampshire.

Charlier, Roger H., and Albert M. Gessman. "Perennial Atlantis." *Sea Frontiers* 18, part 1 (January–March 1972); part 2 (March–April 1972): 40–49, 76–85.

Cox, Allan, G. Brent Dalyrymple, and Richard R. Doell. "Reversals of the Earth's Magnetic Field." *Scientific American* 216 (February 1967): 44–54.

Cox, Allan. "Geomagnetic Reversals." *Science* 169 (January, 1969): 237–44.

Cox, Allan, J. Hillhouse, and M. Fuller. "Paleomagnetic Records of Polarity Transitions, Excursions and Secular Variation," *Transactions of the American Geophysical Union* 13 (1975): 169–185.

Cruxent, Jose M., and Irving Rouse. "Early Man in the West Indies." *Scientific American* 221 (November 1969): 42–52.

Dow, J. W. "Astronomical Orientation at Teotihuacán: A Case Study in Astroarchaeology." *American Antiquity* 32 (1967): 326–34.

Eddy, John A. "Astronomical Alignment of the Big Horn Medicine Wheel." *Science* 184 (June 7, 1974): 1035–43.

Emiliani, Cesare. "A Magnificent Revolution." *Sea Frontiers* 18 (November–December 1972): 357–72.

Fairbridge, Rhodes W. "Global Climate Change During the 13,500 B.P. Gothenburg Geomagnetic Excursion." *Nature* 265 (February 3, 1977): 430–31.

Gifford, John. "The Bimini 'Cyclopean' Complex." *International Journal of Nautical Archaeology and Underwater Exploration.* 2 (1973): 2.

———. "A Description of the Geology of the Bimini Islands, Bahamas." Unpublished master's thesis, University of Miami, Miami, Florida (June 1973).

Glass, Billy P., and Bruce C. Heezen, "Tektites and Geomagnetic Reversals." *Science American* 217 (July 1967): 32–38.

Goodman, Jeffery D. "Psychic Archaeology: Methodology of Empirical Evidence." Paper presented at symposium, Parapsychology and Anthropology. 73rd annual meeting of the American Anthropological Association, Mexico City, November 1974.

Greenman, E. F. "The Upper Paleolithic and the New World." *Current Anthropology* 4 (1963): 41–91; 5 (1964): 321–24.

Harrison, W. "Atlantis Undiscovered—Bimini, Bahamas." *Nature* 230 (April 1971): 287–88.

Harwood, J. M., and S. R. C. Malin. "Present Trends in the Earth's Magnetic Field." *Nature* 259 (February 1976): 469–71.

Hays, James D. "Faunal Extinctions and Reversals of the Earth's Magnetic Field." *Geological Society of America Bulletin* 82 (September 1971): 2433–47.

Hays, J. D., John Imbrie, and N. J. Shackleton. "Variations on Earth's Orbit: Pacemaker of the Ice Ages." *Science* 194 (December 1976): 1121–32.

Heirtzler, J. R. "Project FAMOUS—Man's First Voyages

Down to the mid-Atlantic Ridge." *National Geographic* 147 (May 1975): 586–603.

Hoffer, William. "A Magic Ratio Recurs Throughout Nature." *Smithsonian* 6 (December 1975): 111–24.

Hsü, Kenneth J. "When the Mediterranean Dried Up." *Scientific American* 227 (December 1972): 26–36.

Kennett, James P., and N. D. Watkins. "Geomagnetic Polarity Change, Volcanic Maxima and Faunal Extinctions in the South Pacific." *Nature* 227 (August 1970): 930–34. Deep-sea sedimentary cores from Antarctic-Pacific waters show that some volcanic maxima occurred when the geomagnetic polarity was changing. Upper mantle activity and geomagnetic polarity change may, therefore, be related. Coincidences of faunal extinctions and geomagnetic change may be explained by corresponding volcanically induced change.

Knight, J. W., and P. A. Sturrock. "Solar Activity, Geomagnetic Field and Terrestrial Weather." *Nature* 264 (November 1976): 239–40.

Kolbe, R. W. "Freshwater Diatoms from Atlantic Deep Sea Sediments." *Science* 126 (November 1957): 1053–56.

Kopper, John S., and Stavros Papamarinopoulos. "Human Evolution and Geomagnetism," *Journal of Field Archaeology* 5 (Winter 1978): 443–52.

Kornicker, L. S. "Bahamian Limestone Crusts." *Transactions Gulf Coast Association of Geological Societies* 8 (1958): 167–80.

Lear, John. "The Volcano That Shaped the Western World." *Saturday Review* (November 1966): 57–66.

———. "Were Comets the Midwives at the Birth of Man?" *Saturday Review* (November 1966): 57–66.

Malaise, Rene. "Ocean Bottom Investigations and Their Bearings on Geology." *Geologiska Foreningens I* Stockholm Forhandlinger (March–April 1957): 195–224.

MacNeish, R. S. "Early Man in the New World." *American Scientist* 64 (1976): 316–27.

Marsh, Jacquelyne. "The *Mary Rose:* A Relic of History." *Sea Frontiers* 18 (November–December 1972): 322–28.

McLean, R. F. "Origin and Development of Ridge-Furrow Systems in Beachrock on Barbados, West Indies." *Marine Geology* 5 (1967): 181–93.

Milliman, John D., and K. O. Emery. "Sea Levels During the Past 35,000 Years." *Science* 162 (December 1968): 1121–23.

Morner, Nils-Axel, and Johan P. Lanser. "Gothenburg Magnetic 'Flip.' " *Nature* 251 (October 1974): 408–9.

Newell, Norman D., *et al.* "Organism Communities and Bottom Facies: Great Bahama Bank." *American Museum of Natural History* bulletin 117 (1959): 177–288.

Rebikoff, Dimitri. "Precision Underwater Photomosaic Techniques for Archaeological Mapping; Interim Experiment on the Bimini 'Cyclopean' Complex." *International Journal of Nautical Archaeology and Underwater Exploration* 1 (1972): 184–86.

Reid, George C., and I. S. A. Isaksen. "Marine Extinctions at the Time of Geomagnetic Reversal—690,000 B.P." *Nature* (January 1976).

Richards, Douglas G. "Poseidia '76: A Progress Report." *ARE Journal* 12 (May 1977): pp. 95–104.

Simpson, J. F. "Evolutionary Pulsations and Geomagnetic Polarity." *Geological Society of America Bulletin* 77 (February 1966): 197–204.

Singh, Surenda. "Geomagnetic Activity and Microearthquakes." *Geological Society of America Bulletin* 68 (1978): 1533–35.

Stoddart, D. R. "Three Caribbean Atolls: Turneff Islands, Lighthouse Reef, and Glover's Reef, British Honduras." *Atoll Research Bulletin* 87 (1962): 1–151.

Sykes, Egerton. "The Bermuda Mystery." *Atlantis* 28 (June 1975), the entire issue.

Valentine, J. Manson. "Archaeological Enigmas of Florida and the Western Bahamas." *Muse News* (Miami Museum of Science) 1 (June 1969): 26–29, 41–47.

———. "Culture Pattern Seen." *Muse News* 4 (April 1973), 314–15, 331–34.

INDEX